Applied Science Review

Biology

Applied Science Review™

Biology

Kenneth Zwolski, RN, MS, MA, EdD
Associate Professor
College of New Rochelle
School of Nursing
New Rochelle, N.Y.

Springhouse Corporation
Springhouse, Pennsylvania

Staff

EXECUTIVE DIRECTOR, EDITORIAL
Stanley Loeb

DIRECTOR OF TRADE AND TEXTBOOKS
Minnie B. Rose, RN, BSN, MEd

ART DIRECTOR
John Hubbard

CLINICAL CONSULTANT
Maryann Foley, RN, BSN

EDITORS
Diane Labus, Karen L. Zimmermann

COPY EDITORS
Nancy Papsin, Pamela Wingrod

DESIGNERS
Stephanie Peters (associate art director),
Matie Patterson (senior designer)

COVER ILLUSTRATION
Scott Thorn Barrows

ILLUSTRATORS
Jacalyn Facciolo, Jean Gardner, Robert Neumann

MANUFACTURING
Deborah Meiris (manager), Anna Brindisi,
T.A. Landis

Printed in the United States of America.
ASR1-010792

Library of Congress Cataloging-in-Publication Data
Zwolski, Kenneth.
Biology / Kenneth Zwolski.
p. cm.—(Applied science review)
Includes bibliographical references and index.
1. Biology. I. Title. II. Series
QH308.2.Z96 1993
574—dc20 92-16070
CIP

ISBN 0-87434-453-0

Contents

Advisory Board

Reviewers

Acknowledgments and Dedication

Many people deserve to share in the credit for this book. I would like to acknowledge my gratitude to the reviewers who insured the accuracy of the material and to all the editors and artists at Springhouse who have worked diligently to make the project a success. But most especially, I would like to thank Maryann Foley, who has patiently worked with me since the inception of this book and whose feedback and suggestions have been my guiding force throughout.

To Joann, Audrey, and Alexandra. For all the joy.

Preface

This book is one in a series designed to help students learn and study scientific concepts and essential information covered in core science subjects. Each book offers a comprehensive overview of a scientific subject as taught at the college or university level and features numerous illustrations and charts to enhance learning and studying. Each chapter includes a list of objectives, a detailed outline covering a course topic, and assorted study activities. A glossary appears at the end of each book; terms that appear in the glossary are highlighted throughout the book in boldface italic type.

Biology provides conceptual and factual information on the various topics covered in most biology courses and textbooks, and focuses on helping students to understand:

- the nature of science and biology
- the chemical components of all living organisms
- cell structure and function
- mechanisms for reproduction and growth
- regulation and control mechanisms in plants and animals
- principles of inheritance
- evolution and ecology
- similarities and differences among various living organisms, including viruses, monera, protista, fungi, plants, and animals
- the organization and function of body systems in vertebrates.

1

Overview of Science and Biology

Objectives

After studying this chapter, the reader should be able to:
- Define science.
- Describe the four steps of the scientific method.
- Define biology.
- Identify at least 10 of the specialized branches of biology.
- Outline each of the unifying themes that characterize biology.

I. Science

A. General information

1. Science, derived from the Latin word *scire* ("to know"), is knowledge about the general laws and operation of the universe obtained through research and analysis
2. Physical science encompasses various fields of study, including biology, chemistry, and physics
3. Scientists research and analyze information to understand it and solve problems
4. The *scientific method* is the primary tool by which scientists gather and analyze information

B. The scientific method

1. A systematic, logical procedure for obtaining knowledge, the scientific method consists of four steps: observation, hypothesis formation, experimentation, and interpretation
2. *Observation* is the act by which the scientist recognizes and notes a specific phenomenon that can be measured objectively
    a. During this step, the scientist documents all relevant data, including a detailed description of the phenomenon and its components (variables)
    b. Observations must be limited to those the scientist can observe directly or indirectly, with or without an instrument
    c. A phenomenon that cannot be quantified objectively is considered unsuitable for further scientific study
    d. After collecting the data, the scientist generalizes about the observations and hypothesizes about the mechanisms involved
3. A *hypothesis* is an educated guess about the possible relationship between two or more variables

a. The purpose of a hypothesis is to develop a proposition (a formal statement that asserts or explains a relationship between the variables) that can be tested experimentally
b. A well-worded proposition indicates clearly how the variables might be related, such as "A is bigger than B because..."
4. *Experimentation* is a controlled operation for testing the hypothesis
a. An experiment requires the running of two parallel tests that are identical in all but one respect: in one test, the variable of interest is applied; in the other, it is not
(1) The test involving the applied variable is called the *experimental*
(2) The test in which the variable is not applied is called the *control;* the control serves as the scientist's frame of reference during the experiment
b. Any difference between the two tests is evidence of the variable's effect
5. *Interpretation,* the final step of the scientific method, critically examines the results of the experiment
a. Based on this critical analysis, the original hypothesis may be accepted, rejected, or modified
b. Scientific knowledge is gained through the interpretation step

II. Biology

A. General information

1. Biology is a broad branch of science that is concerned with all living things
2. It encompasses many specialized branches that focus on singular aspects of life or living things
a. *Zoology* is the study of animals
b. *Botany* is the study of plants
c. *Mycology* is the study of fungi
d. *Biochemistry* is the study of the chemical compounds and processes in living things
e. *Cytology* is the study of cellular structure
f. *Embryology* is the study of growth and development from conception to birth
g. *Physiology* is the study of the normal function of a living organism
h. *Anatomy* is the study of the structure of living things
i. *Genetics* is the study of heredity and the mechanisms by which traits are inherited
j. ***Ecology*** is the study of the interrelationships of organisms with their environment
k. *Microbiology* is the study of microorganisms
l. *Histology* is the study of tissues
m. *Cell biology* is the study of cells
n. *Reproductive biology* is the study of reproductive structures and processes
o. ***Taxonomy*** is the study of the classification of living things based on their relationships with each other
p. *Molecular biology* is the study of the molecular structure of living organisms
q. *Aquatic biology* is the study of aquatic organisms

3. Specialized branches may be further subdivided or defined; for example, *population ecology,* a subdivision of ecology, is the study of how populations of living organisms interrelate
4. Specialized branches of biology may overlap (for example, aquatic biology overlaps zoology because both study animals; however, aquatic biologists do not study terrestrial animals)
5. Biologists sometimes rely on the knowledge of other scientists to explain a phenomenon (for example, Charles Darwin drew on knowledge from paleontology [the study of fossils] and geology [the study of the earth's history as recorded in rocks] to explain how living things change; he combined facts from these sciences with his own observations of nature to arrive at the theory of evolution)

B. Major themes in biology

1. Biologists have identified certain unifying themes common to all living organisms, including ***hierarchy*** of organization, common ancestry, and adaptability to a changing environment
2. Living organisms are organized according to a hierarchy, wherein each level builds upon the previous one yet retains the basic properties of life
   a. At the most basic level, organisms are composed of molecules in patterns that form the basic structure of cells; in more advanced organisms, cells are organized into tissues, tissues into organs, and organs into complex systems and populations
   b. The organizational hierarchy is based on the tenet that the whole is greater than the sum of its parts
3. Although all organisms are composed of the same basic elements, they are highly diverse; biologists have identified about 1.7 million different species, from single-celled organisms to plants and animals
4. Despite their great diversity, all living things share the same universal genetic code by which biological information is passed from one generation to the next; consequently, all organisms are governed by the same rules of inheritance
   a. Reproduction is the mechanism by which all organisms pass on biological information
   b. Some organisms reproduce exact duplicates of themselves; others fuse with a member of the opposite sex, producing offspring that are similar to both parents but with different traits
   c. Reproduction is essential for the survival of the species and the continuation of each organism's biological information
5. According to the theory of ***evolution,*** all living organisms have descended from a common ancestor and have transformed from simpler to more complex forms
6. Species are not fixed but change over time in response to a changing environment
   a. All organisms constantly interact with their environment and with other organisms
   b. Living organisms respond and adjust to their environment while maintaining an internal equilibrium (homeostasis)
7. Every structure in a living organism is perfected to perform a vital function

Study Activities

1. Organism A changes color in the presence of organism B. Form a hypothesis to explain this phenomenon, and describe how the hypothesis could be tested experimentally.
2. List and define five of the specialized branches of biology. Identify how these branches overlap.
3. Give three examples to illustrate each of the following unifying themes in biology: the hierarchical organization of life, the diversity and unity of life, the ability of organisms to adjust to their environment, the interrelationship of organisms to each other and their environment, the correlation of structure with function, and the transformation of life over time.

2

The Chemistry of Life

Objectives

After studying this chapter, the reader should be able to:

- Describe the structure of the atom.
- Discuss the four major classes of organic compounds that occur in living matter.
- Discuss the role of water in organic reactions.
- Identify the steps involved in cellular respiration.
- Describe the process and purpose of photosynthesis.
- Write the overall chemical reactions that occur in both aerobic respiration and photosynthesis.

I. The Building Blocks of Life

A. General information

1. All living things are composed of matter, a material substance that has weight and occupies space
2. Elements, the building blocks of all matter, are made up of atoms having the same basic properties
3. When two or more elements join, they form substances known as compounds

B. Atoms

1. *Atoms* are the fundamental units of all matter
2. Each atom consists of a central nucleus surrounded by a characteristic number of moving, negatively charged particles (electrons)
   a. The nucleus of an atom is about $\frac{5}{1,000}$ the size of the atom, yet it accounts for nearly all the atom's weight
   b. The nucleus consists primarily of two kinds of particles: *protons,* positively charged particles, and *neutrons,* particles with no charge
3. *Electrons* are negatively charged particles that are arranged around the nucleus of an atom in energy levels or shells
4. Each electron shell is composed of orbitals —areas of three-dimensional space in which an electron will be found about 90% of the time; no more than two electrons can occupy the same orbital at any given time
   a. The first electron shell (1s) contains only one spherical orbital that can hold up to two electrons
   b. The second energy shell contains one spherical orbital (2s) and three dumbbell-shaped orbitals (2px, 2py, and 2pz); each orbital can hold up to two electrons for a total of eight electrons

   c. Higher electron shells (those further from the nucleus) can have additional orbitals with more complex shapes to accommodate many electrons
   d. The outermost electron shell of any atom never has more than four orbitals (one "s" orbital and three "p" orbitals) and therefore never contains more than eight electrons
5. In its normal state, an atom contains an equal number of protons and electrons, resulting in an electrical balance between the positive and negative charges

C. Elements

1. An *element* is a substance composed of similar atoms
2. Elements are distinguished from each other by the number of protons contained in their nuclei; for example, the element carbon comprises all atoms with six protons in the nucleus, whereas the element oxygen is made up of all atoms with eight protons in the nucleus
3. Each element possesses unique properties, such as its normal physical state (solid, liquid, or gas) and its potential to react with other substances, that distinguishes it from all other elements
4. Scientists so far have identified 109 elements —92 naturally occurring elements and 17 artificially produced ones
5. All atoms of the same element have the same atomic numbers but may have different atomic weights
   a. The *atomic number* equals the number of protons in the nucleus of each atom; it is indicated by a subscript to the left of the element's symbol (for example, the element carbon [C], which has an atomic number of 6, is written as ${}_6C$)
   b. The *atomic weight* equals the number of protons plus the number of neutrons in each atom; it is indicated by a superscript to the left of the element's symbol (for example, carbon atoms normally have an atomic weight of 12 [six neutrons and six protons], written as ${}^{12}C$)
6. When all the atoms have the same number of protons but a different number of neutrons, the element is referred to as an *isotope*
7. Some isotopes occur naturally, whereas others can be created by chemists under laboratory conditions
8. Hydrogen has three isotopes: hydrogen (the most common), which has one proton and no neutron (${}^1H$); deuterium, which has one proton and one neutron (${}^2H$); and tritium, which has one proton and two neutrons (${}^3H$)

D. Molecules and compounds

1. A *molecule* is the combination of two or more similar or different atoms
   a. It is the smallest possible unit of a substance that still retains the properties of that substance
   b. Molecules are capable of stable, independent existence
2. A *compound* is the combination of molecules in a definite proportion by weight
   a. It always contains atoms of more than one element and can be reduced to its elements
   b. A compound has properties different from those of the elements that compose it
   c. Commonly occurring compounds include H_2O and table salt (NaCl)

II. Properties of Chemical Substances

A. General information

1. The chemical behavior of an atom (its potential reactivity with other atoms) is determined by the distribution of electrons in its outermost shell
2. The outermost shell, or *valence shell,* may contain a complete or an incomplete set of electrons
   a. An atom with a complete valence shell (one that contains all of its eight electrons) is considered chemically stable —that is, it is unreactive, or inert
   b. An atom with an incomplete valence shell (one having fewer than eight electrons) is chemically unstable and will interact with other atoms to form stable molecules or compounds
3. Atoms form stable compounds by bonding with other atoms; such bonding enables the atoms to either share or transfer electrons so that each atom can complete its valence shell
4. ***Acids*** and ***bases*** are special compounds that alter the hydrogen ion concentration of a solution; the hydrogen ion concentration determines the acidity or basicity (alkalinity) of a solution

B. Chemical bonds

1. Two or more atoms are held together through a force of attraction called a *chemical bond*
2. Chemical bonds can store energy in the form of potential energy; the position the electrons assume in forming a bond determines the amount of energy stored
3. A chemical bond may be one of three types: covalent, polar covalent, or ionic
   a. In a *covalent* bond, the atoms share electrons; for example, in the hydrogen molecule (H_2), two atoms of the element hydrogen share electrons to form a single molecule
   b. In a *polar covalent* bond, the atoms share electrons but the electrons are held more tightly to one of the atoms, resulting in a slightly negative charge at one end of the molecule and a slightly positive charge at the other end; for example, in H_2O, the electrons are attracted more strongly to the oxygen than to the hydrogen, resulting in a relatively negative charge at the oxygen end and a relatively positive charge at the hydrogen end
   c. In an *ionic* bond, the atoms completely transfer electrons from one to another—that is, one atom gains an electron while another loses an electron, enabling each atom to complete its valence shell; for example, in the compound sodium chloride (NaCl), the sodium (Na) loses an electron whereas the chlorine (Cl) gains an electron
4. An atom that forms an ionic bond is known as an *ion;* such an atom will have a net negative charge if it gains an electron and a net positive charge if it loses an electron
   a. A *cation* is an ion with a net positive charge (for example, Na^+)
   b. An *anion* is an ion with a net negative charge (for example, Cl^-)

C. Chemical reactions

1. Chemical reactions occur when chemical bonds are formed or broken
2. These reactions may be one of two types: endergonic or exergonic

a. An *endergonic* chemical reaction requires the input of energy from the surrounding area, usually from heat or light

b. An *exergonic* chemical reaction releases energy into the surrounding area

3. *Ionization* is a special type of chemical reaction in which ionic bonds are broken and the ions separate; for example, if the compound NaCl is placed in water, the ionic bonds break and the Na^+ and Cl^- ions dissociate from each other
4. Water plays a key role in certain chemical reactions; in some cases, water is released as a result of the reaction, whereas in other cases it is required for the reaction to take place

D. Acids and bases

1. A water molecule is capable of dissociating into a hydrogen ion (H^+) and a hydroxide ion (OH^-)

$$H_2O \leftrightarrows H^+ + OH^-$$

2. Because this dissociation produces one hydrogen ion for every hydroxide ion, both ions are normally equal in pure water; however, special compounds placed in water can either increase or decrease the hydrogen ion concentration, thereby altering the normal ionic balance

   a. An *acid* is a substance that increases the hydrogen ion concentration of a solution (for example, when hydrochloric acid [HCl] is dissolved in water, it dissociates into H^+ and Cl^- ions, increasing the concentration of hydrogen ions in the water)

   b. A *base* is a substance that decreases the hydrogen ion concentration of a solution (for example, when sodium hydroxide [NaOH] is dissolved in water, it dissociates into Na^+ and OH^- ions; the OH^- ions then combine with H^+ ions to form H_2O, thereby decreasing the concentration of hydrogen ions in the water)

3. The concentration of hydrogen ions in a solution determines the acidity or basicity (alkalinity) of the solution
4. The ***pH scale*** is the tool by which scientists measure the H^+ ion concentration of a solution to determine its acidity or alkalinity

   a. A pH of 7.0 is neutral (neither acidic nor alkaline)

   b. A pH less than 7.0 is acidic; a pH greater than 7.0, alkaline

5. Living organisms are highly sensitive to changes in pH levels and cannot withstand wide variations in pH concentrations
6. ***Buffers*** are substances that minimize changes in the concentration of H^+ and OH^- ions when acids and bases are introduced into living tissues
7. The pH of various tissues in the human body ranges from 3.0 to 8.5; the pH of human blood is maintained by buffers at approximately 7.4

III. Organic Compounds

A. General information

1. Organic compounds consist primarily of carbon and hydrogen
2. These compounds, which are the most prevalent compounds found in living matter, include carbohydrates, lipids, proteins, and nucleic acids
3. They tend to be large, complex compounds that form covalent bonds

B. Carbohydrates

1. *Carbohydrates* are compounds containing the elements carbon (C), hydrogen (H), and oxygen (O)
2. The hydrogen and oxygen occur in a 2:1 ratio, the same as in water
3. *Monosaccharides* are the simplest carbohydrates
   a. Their formula is usually some multiple of CH_2O
   b. Glucose, whose formula is $C_6H_{12}O_6$, is the most common and most important monosaccharide because it is the source of energy for most organisms
   c. Other monosaccharides include fructose and galactose
4. *Disaccharides* are two monosaccharides linked together
   a. Maltose is a disaccharide comprised of two glucose molecules linked together
   b. Other disaccharides include sucrose (table sugar), which is glucose linked to fructose, and lactose (milk sugar), which is glucose linked to galactose
5. *Polysaccharides* consist of hundreds of glucose molecules bound together in complex chains
   a. Three major polysaccharides include starch, glycogen, and cellulose
   b. Starch is the major storage form of glucose in plants; glycogen is the major storage form of glucose in animals; and cellulose is the building material found in the cell walls of most plants

C. Lipids

1. *Lipids*—water-insoluble compounds containing carbon and hydrogen but less oxygen than that found in carbohydrates—cushion vital organs, provide insulation, store food, and speed up the conduction of nerve impulses
2. The structural building block of a lipid is the *fatty acid*
   a. Fatty acids consist of a chain of carbon and hydrogen atoms attached to a carboxyl group (COOH)
   b. The carbon chain of a fatty acid can vary from 4 to 24 atoms
3. The three most important types of lipids are fats (saturated and unsaturated), phospholipids, and steroids
   a. *Fats* contain three fatty acids combined with a glycerol (sugar alcohol); they may be solid or liquid at room temperature
      (1) *Saturated fats* are lipids whose fatty acids contain the maximum number of hydrogen atoms attached to each carbon (two or three, depending on the position of carbon in the carbon chain); they tend to be solid at room temperature
      (2) *Unsaturated fats* are lipids whose fatty acids do not contain the maximum number of hydrogen atoms attached to each carbon; they tend to be liquid at room temperature
   b. *Phospholipids* contain phosphoric acid and a nitrogenous compound in addition to glycerol and fatty acids
      (1) The fatty acids in a phospholipid are nonpolar and therefore insoluble in water; the phosphate group is electrically charged (polar) and therefore soluble in water
      (2) Because one end of the phospholipid is soluble in water and the other end is not, the phospholipid is a crucial component of many cellular membranes

c. *Steroids,* although structurally different from the other lipids, are classified as such because they are insoluble in water; many hormones are made up of steroids

D. Proteins

1. *Proteins,* sometimes called polypeptides, are polymers (large molecules) composed of as many as 3,000 amino acids bonded together
2. An *amino acid* is a compound comprised of carbon, hydrogen, an amino group (NH_2), a carboxyl group, and a side group designated "R," which distinguishes one amino acid from another
    a. The 20 types of amino acids can be combined in various ways to form many different types of proteins
    b. The bond between amino acids, called a *peptide bond,* links the carboxyl group of one amino acid to the amino group of another
    c. When combined to form a protein, amino acids can be arranged in a straight chain, a twisted chain (helix), or a globule
3. In living tissues, proteins provide support; transport substances within the cell or body; form enzymes, antibodies, and hormones; and promote contractility of tissue through the actions of two special proteins, actin and myosin

E. Enzymes

1. *Enzymes* are special proteins that regulate the rate of nearly every chemical reaction; they provide energy to cells, help build new cells, and aid in digestion
2. Each cell contains thousands of enzymes, and each chemical reaction is regulated by a specific enzyme
3. Because enzymes remain unchanged, they can be used over and over again
4. Enzymes commonly work together with a co-enzyme
    a. A *co-enzyme,* a small, nonprotein compound, functions only in the presence of an enzyme
    b. Examples of co-enzymes include the B complex vitamins (such as thiamine, riboflavin, and niacin)
5. Although scientists do not understand fully how enzymes work, many adhere to the lock-and-key model of enzyme activity (see *The Lock-and-Key Model of Enzyme Activity*)
6. Various factors affect enzyme activity, including pH, temperature, and the concentration of enzymes and substrates (the substances on which enzymes work)
    a. An optimal pH level is necessary for enzyme activity to occur; in humans, the pH level must be about 7.0 for enzymes to work, but in some reactions (such as digestion), the ideal pH level is closer to 1.5
    b. In humans, most reactions work best when the body temperature is 98.6° F (37° C)
        (1) Any decrease in temperature deceases the rate of enzyme activity
        (2) Any increase in temperature increases the rate of enzyme activity; however, enzymes begin to become denatured—undergo a change in physical properties—once a maximum temperature is reached (the highest tolerable temperature in humans is about 104° F [40° C]; in bacteria, 158° F [70° C])

The Lock-and-Key Model of Enzyme Activity

According to the lock-and-key model of enzyme activity, an enzyme is capable of binding to a specific part of a substrate because of its unique shape, which complements exactly the shape of the substrate. The enzyme holds the substrate in place (the point at which the reaction takes place is called the active site) and catalyzes the reaction — in this case, breaking the bond between A and B.

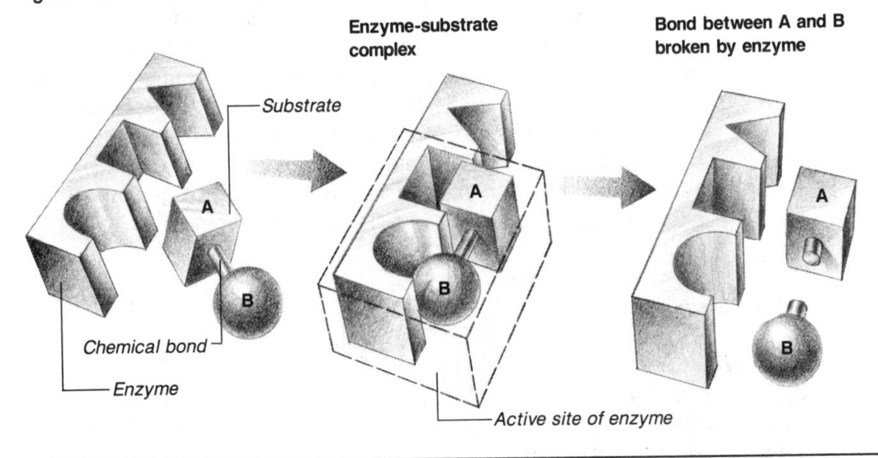

c. The concentration of the enzyme and substrate affects the rate and constancy of chemical reactions
   (1) If the amount of enzyme is increased and the substrate remains constant, the reaction rate increases to a certain point, after which it remains constant
   (2) If the amount of substrate is increased and the enzyme remains constant, the reaction rate increases to a point at which every available enzyme molecule is actively involved in the reaction

F. Nucleic acids

1. *Nucleic acids* are large organic molecules comprised of carbon, hydrogen, oxygen, nitrogen, and phosphorus
2. The only organic compounds capable of replicating themselves, nucleic acids are involved in the storage and transmission of energy and cellular information
3. The structural building block of a nucleic acid is the *nucleotide,* which is comprised of a phosphate group (PO_4), a five-carbon sugar (ribose or deoxyribose), and a nitrogen base, either a purine (adenine or guanine) or a pyrimidine (cytosine, thymine, or uracil)
4. Nucleic acids include deoxyribonucleic acid (DNA), ribonucleic acid (RNA), cyclic adenosine monophosphate (cAMP), guanosine triphosphate (GTP), and adenosine triphosphate (ATP)
   a. DNA, which can replicate itself, stores information that directs the cell's activities
   b. RNA, which is stored in DNA, is involved in the transcription and translation of this information

(1) *Transcription* is the transfer of information from a DNA molecule to an RNA molecule; it occurs in the nucleus
(2) *Translation* is the transfer of information from an RNA molecule to a polypeptide; it occurs in the cytoplasm

c. cAMP and GTP act as chemical messengers, relaying information that affects biological responses

d. ATP, the principal energy-carrying molecule of all cells, stores and releases energy for cellular functions by forming and breaking chemical bonds
(1) Although ATP is necessary to generate energy for many of the basic processes of living organisms (including movement, heat production, and excretion of wastes), it is found in only limited quantities in cells and must be continuously replenished
(2) The most common and efficient way in which living organisms replenish ATP is through a process called cellular respiration

IV. Cellular Respiration

A. General information

1. *Cellular respiration,* a complex metabolic process by which energy is released from the breakdown of food molecules (glucose), provides cells with necessary fuel to carry out basic functions
2. It involves a series of chemical reactions and is the principal way in which ATP—the energy storehouse of cells—is replenished
3. Cellular respiration occurs in animal and plant cells
   a. In animal cells, glucose is obtained from food, either by direct ingestion of glucose or the conversion of proteins and lipids into glucose
   b. In plant cells, glucose is manufactured through photosynthesis
4. Glucose ($C_6H_{12}O_6$) may be broken down either *aerobically* (using oxygen) or *anaerobically* (without oxygen)

B. Aerobic respiration

1. Aerobic respiration occurs in three phases: glycolysis, the Krebs cycle, and the electron transport chain
2. *Glycolysis* involves a series of chemical reactions through which glucose molecules are converted to ATP in the absence of oxygen; this process can take place in either the mitochondria or cytoplasm of the cell
   a. The 6-carbon glucose molecule is converted into two 3-carbon molecules of pyruvic acid
   b. Hydrogen is then removed from the molecules, and the energy released is captured and stored in four ATP molecules (this energy is used later to add phosphate to adenosine diphosphate to form ATP)
   c. The energy stored in two ATP molecules is required to begin the chemical reaction, resulting in a net gain of two ATP molecules
3. In the *Krebs cycle,* or citric acid cycle, pyruvic acid is degraded in the presence of oxygen
   a. Intermediate compounds formed during the Krebs cycle include oxaloacetic acid, citric acid, succinic acid, and ketoglutaric acid

b. Additional hydrogen is removed during the Krebs cycle, and some of the energy released by this process is used to generate two more molecules of ATP

4. In the *electron transport chain,* the hydrogen electrons removed during the previous two phases are passed first to a hydrogen acceptor (nicotinamide-adenine dinucleotide [NAD]), and then to a class of compounds known as cytochromes
   a. As the hydrogen electrons are transported from one hydrogen acceptor to the next, energy is released and stored in ATP molecules
   b. Oxygen, the final hydrogen acceptor in the chain, then combines with the hydrogens to form H_2O; carbon dioxide (CO_2) also is generated at this stage
   c. The electron transport chain produces 32 ATP molecules
5. In aerobic respiration, a single molecule of glucose yields a total of 36 energy-rich ATP molecules
6. The overall reaction for aerobic respiration can be written as follows:

$$C_6H_{12}O_6 + 6O_2 + 36ADP + 36P \rightarrow 6CO_2 + 6H_2O + 36ATP$$

C. Anaerobic respiration

1. Anaerobic respiration, also called fermentation, takes place in the cell's cytoplasm
2. It occurs without oxygen and involves only glycolysis, not the Krebs cycle or the electron transport chain
3. Because anaerobic respiration involves fewer steps than aerobic respiration, it produces fewer (only two) ATP molecules and less energy
   a. Glycolysis occurs the same as in aerobic respiration; however, the pyruvic acid generated from the first phase is changed to lactic acid or ethyl alcohol and carbon dioxide
   b. No new ATP molecules are created by the conversion of pyruvic acid to lactic acid or ethyl alcohol and carbon dioxide

V. Photosynthesis

A. General information

1. ***Photosynthesis*** is the process by which ***autotrophs***—organisms capable of making their own food from organic material—manufacture glucose
2. All animals depend directly or indirectly on autotrophs for food and energy
3. Photosynthesis differs from cellular respiration in many ways; it requires chlorophyll, light reactions, and carbon dioxide to take place and yields glucose and water, which can be used later by the plant (see *Comparing Photosynthesis with Respiration,* page 14, for a brief summary of the mechanisms involved in these two processes)
4. The overall reaction for photosynthesis can be written as follows:

$$12H_2O + 6CO_2 \rightarrow C_6H_{12}O_6 + 6H_2O + 6O_2$$

B. Chlorophyll

1. A mixture of pigments, ***chlorophyll*** is comprised primarily of chlorophyll A and B (green pigments) but also may contain carotenes and xanthophylls (yellow pigments), phycobilins, phycoerythrins, and phycocyanins; carbon, hydrogen, oxygen, nitrogen, and magnesium also are found in chlorophyll

Comparing Photosynthesis with Respiration

EVENT	PHOTOSYNTHESIS	RESPIRATION
When it occurs	Day only	Day and night
What is taken in	$CO_2 + H_2O$	O_2
What is given off	O_2	$CO_2 + H_2O$
Where it takes place	Plant cells containing chlorophyll	All cells (plant and animal)
What happens to glucose	Synthesized	Broken down
What happens to energy	Energy from sunlight is used, stored in glucose or starch	Energy from glucose is used, stored in ATP
Overall reaction	$12H_2O + 6CO_2 \rightarrow C_6H_{12}O_6 + 6H_2O + 6O_2$	$C_6H_{12}O_6 + 6O_2 + 36ADP + 36P \rightarrow 6CO_2 + 6H_2O + 36ATP$

2. Chlorophyll is found in tiny oval bodies called ***chloroplasts;*** these cellular organelles are located in all the green parts of a plant
3. Chlorophyll is formed only in the presence of light (usually sunlight but also artificial light)
4. Certain wavelengths of light, such as the blue-violet and orange-red ends of the spectrum, are best absorbed by the chlorophyll pigment within the chloroplast
   a. Light energy activates chlorophyll's electrons, which absorb and transfer the energy to other chemicals within the chloroplast; this light energy also splits water (a process called *photolysis*)
   b. When depleted of energy, the electrons return to the chlorophyll pigment to become reenergized, thus beginning the cycle anew
   c. The formation of ATP molecules during this process is known as *cyclic photophosphorylation.*

C. Light reaction

1. The *light reaction* is a series of photochemical reactions that converts light energy to chemical energy
2. It occurs in special flattened, membranous stacks within the chloroplast called *thykaloids*
3. In this process, light activates the chlorophyll and splits the individual water molecules that are brought up to the chloroplast from the roots
   a. After a water molecule is split (photolysed), the hydrogen from the water combines with nicotinamide-adenine dinucleotide phosphate (NADP) to form $NADPH_2$, which subsequently is used in a dark reaction
   b. The oxygen from the water is released into the atmosphere
4. The energy from the light is also harnessed to produce ATP (this results from the transfer of some light energy into ADP bonds)

D. Dark reaction

1. The *dark reaction* (also called the *Calvin cycle*), which does not require light yet occurs only during the day, incorporates carbon into organic compounds
2. In this process, enzymes use the energy stored in $NADPH_2$ and ATP (formed during the light reaction) to convert CO_2 (obtained from the atmosphere) into glucose and water
   a. The enzymes needed for the dark reaction are stored in *stroma*—the fluid ground substance of chloroplasts
   b. The glucose and water resulting from the dark reaction can be stored for future use by the plant

Study Activities

1. Diagram an atom with three electron shells, labeling each orbital and identifying the maximum number of electrons in each shell.
2. Write out the formula for a simple carbohydrate, lipid, and protein. Describe what feature distinguishes each compound from the others.
3. Outline the lock-and-key model of enzyme activity, including those factors that affect such activity.
4. Sketch the basic structure of a nucleotide. Label the phosphate group, five-carbon sugar, and nitrogen base.
5. Compare aerobic and anaerobic respiration.
6. Identify which of the end products of photosynthesis result from the light reaction and which result from the dark reaction.

3

Cellular Structure and Function

Objectives

After studying this chapter, the reader should be able to:

- Identify at least 10 basic functions that all living organisms can perform independently.
- Name the three main cellular components and describe their functions.
- Explain the difference between prokaryotic and eukaryotic cells.
- Define differentiation and discuss the normal pattern of differentiation in plants and animals.
- Explain the primary difference between simple plant tissue and complex plant tissue.
- Describe the function of roots, stems, and leaves in a plant.
- Define the four major types of animal tissue found in vertebrates and explain their function.

I. The Cell

A. General information

1. The cell is the basic unit of structure and function in all living things
2. The cell responds to the environment by continuously exchanging materials and energy with its surroundings
3. The average size of a cell is between 5 and 15 microns (a micron is equivalent to 1/1,000 of an inch)
4. All cells are considered alive because they can carry out basic functions
    a. *Movement*—the cell's ability to reposition itself or change shape
    b. *Reproduction*—the cell's ability to divide, producing identical or similar copies of itself
    c. *Ingestion*—the ability to take in food
    d. *Digestion*—the breaking down of food into simpler forms through enzyme action
    e. *Absorption*—the movement of dissolved materials through the cell membrane
    f. *Assimilation*—the conversion of nonliving material into protoplasm
    g. *Synthesis*—the formation of more complex molecules from simpler ones
    h. *Secretion*—the release of functional or useful substances, such as enzymes
    i. *Excretion*—the release or expulsion of superfluous or toxic substances
    j. ***Respiration***—the release of energy from food molecules (with or without oxygen)
    k. *Transport*—the movement of substances within the cell

l. *Regulation*—the ability to self-monitor cellular processes
m. *Irritability*—the capacity to respond to stimuli

B. Classification and function

1. All cells can be classified into one of two categories based on the presence or absence of an enclosed (or true) nucleus
   a. *Eukaryotic* cells contain a clearly defined central area (nucleus) enclosed by a membrane; they occur in two major patterns—plant and animal
   b. *Prokaryotic* cells contain a vaguely defined, unenclosed central area (nuclear area or nucleoid); these cells are the type found in all bacteria and blue-green algae
2. Cells can exist independently or together in specialized groups (tissues, organs, and organ systems)
3. When cells aggregate into more complex structures, they assume specialized functions and lose their ability to survive independently

II. Cellular Components

A. General information

1. All cells consist of three basic components: a central region *(nucleus* or *nuclear area)* that controls cellular activity, a semifluid medium ***(cytoplasm)*** that surrounds the nucleus, and a semipermeable barrier *(plasma membrane)* that encloses the cell
2. Various structures *(organelles)* are contained within the cytoplasm of all cells, each having a specific function
3. Plant cells are differentiated from all other cells by the presence or absence of certain structures, such as a cell wall, chloroplast, and large vacuoles

B. Nucleus

1. Typically located within the central region of the cell, the nucleus (or nuclear area) plays an integral role in cell reproduction
2. The nucleus primarily consists of chromosomes and a nucleolus
   a. ***Chromosomes*** are threadlike structures comprised of genetic material (deoxyribonucleic acid, or DNA) and protein
   b. The *nucleolus,* the most visible structure within the nondividing nucleus, manufactures and controls *ribosomes,* which help to synthesize protein
3. Some cells contain two or more nucleoli in the nucleus
4. Depending on the type of cell, the nucleus may or may not be enclosed by a membrane
   a. In eukaryotic cells, the nucleus is enclosed by a *nuclear membrane,* a double-membrane envelope that regulates the flow of material into and out of the nucleus
   b. In prokaryotic cells, the genetic material is concentrated in a region called the *nucleoid,* which is not separated from the rest of the cell by a membrane

Animal Cell

This illustration shows the most common cellular components of a typical animal cell. Note that the cell, which contains a central region (nucleus) surrounded by various organelles suspended in cytoplasm, is enclosed by a thin plasma membrane.

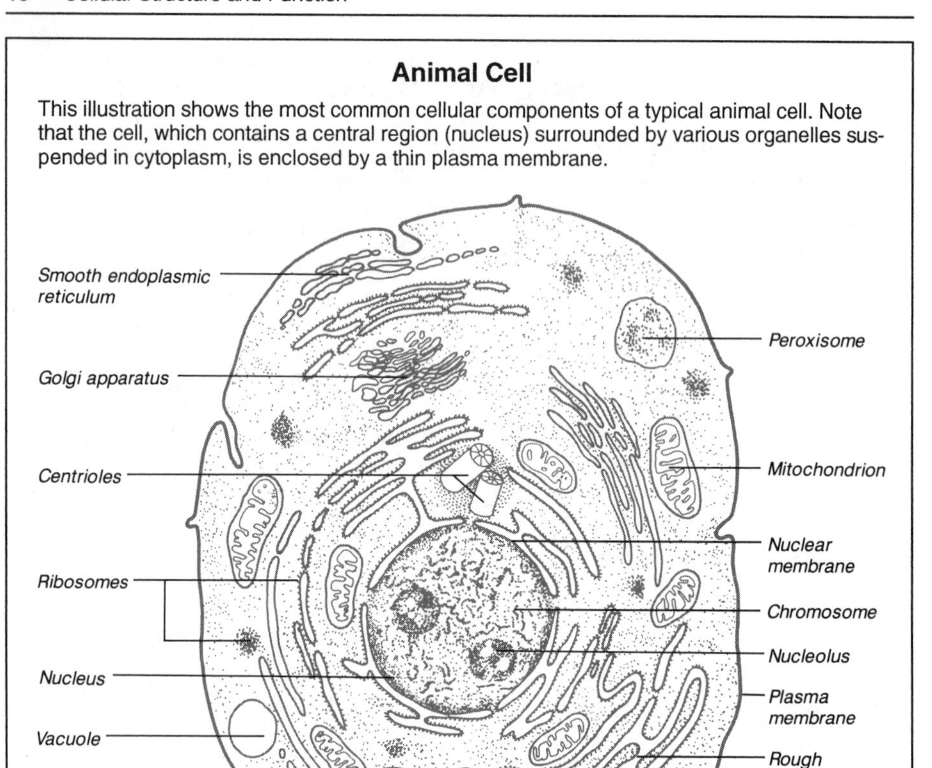

C. Cytoplasm

1. The cytoplasm is the area between the plasma membrane and the nuclear membrane of the cell
2. It consists of *cytosol,* a semifluid, jellylike medium in which organelles of specialized form and function are suspended
3. The *cytoskeleton,* a network of fibers distributed throughout the cytoplasm, helps with cell support and locomotion
   a. *Microtubules,* found in all cells, are straight, hollow rods that typically radiate from an organizing center (a mass located near the nucleus); they serve as strong girders to help support the cell
      (1) Within the microtubule-organizing center of animal cells are two structures called *centrioles* that help with chromosome separation during cell division
      (2) *Cilia* and *flagella* are specialized arrangements of microtubules that provide locomotion for small cells
      (3) They are common in prokaryotic cells and animal cells but rare in plant cells

Plant Cell

This illustration shows the most common cellular components of a typical plant cell. Note that the cell, which contains a nucleus surrounded by various organelles suspended in cytoplasm and a large storage area (vacuole), is encased in a thick cell wall.

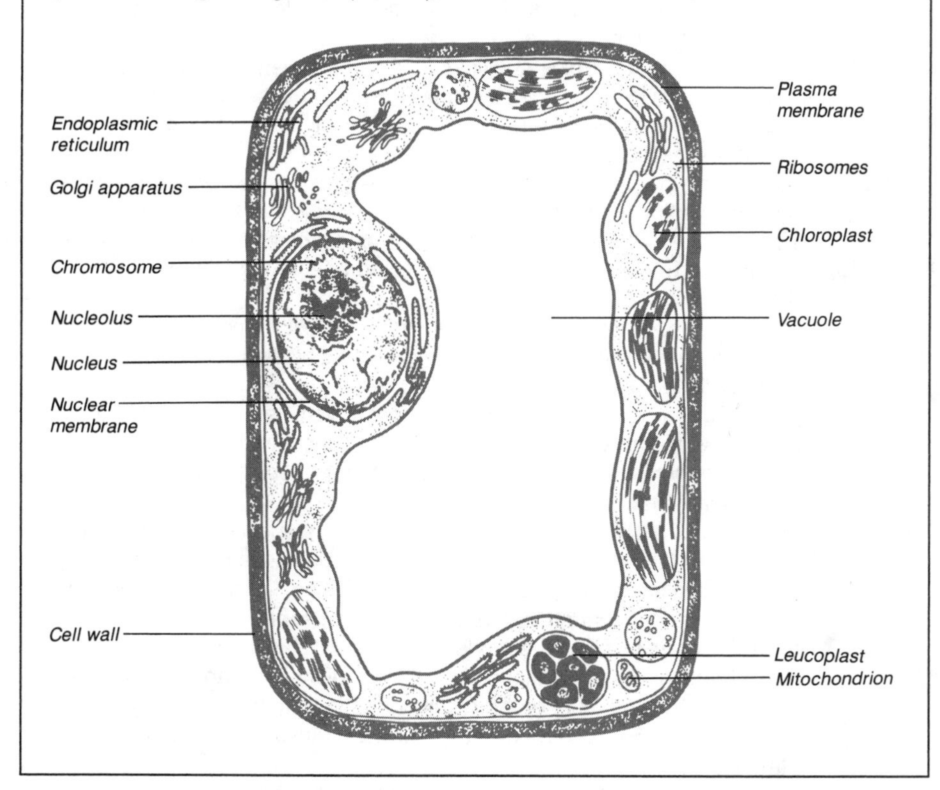

   (4) Cilia are smaller than flagella and outnumber flagella on the surface of the cell
   (5) Both cilia and flagella are anchored in the cell by a basal body that is structurally identical to a centriole

b. *Microfilaments,* found in all cells, are solid rods that provide support and help localize cell contractions
   (1) They also help the cell to lengthen and retract during amoeboid movement
   (2) Microfilaments may assist in a phenomenon in plant cells called ***cytoplasmic streaming,*** in which the entire cytoplasm circulates around the space between the vacuole and the plasma membrane; cytoplasmic streaming speeds up the distribution of materials within the cell

D. Plasma membrane

1. A semipermeable barrier between the internal cell and its external environment, the plasma membrane can be explained by one of two models

a. The *Davson-Danielli model,* developed in 1935, depicts the plasma membrane as similar to a sandwich —two layers of phospholipids situated between two layers of globular proteins
b. The *Fluid Mosaic model,* a more accepted view developed in 1972, depicts the plasma membrane as dispersed and embedded proteins in a phospholipid bilayer that is in a fluid state
2. Primarily composed of proteins and lipids, the plasma membrane regulates the chemical composition of the cell by allowing some substances to pass readily between the cell and the external environment while impeding the entrance or exit of other substances
3. Besides allowing the passage of substances, the plasma membrane also aids in cell-cell recognition, whereby one cell recognizes another by the composition of its plasma membrane, and in adenosine triphosphate (ATP) reactions
4. Substances may move across the plasma membrane through various transport mechanisms
a. *Passive transport* involves the movement of molecules across the plasma membrane without the use of energy, such as by diffusion, osmosis, or carrier-facilitated diffusion
(1) ***Diffusion*** is the molecular movement of particles from an area of greater concentration to one of lesser concentration
(2) ***Osmosis*** is the movement of water across a selectively permeable membrane
(3) ***Carrier-facilitated diffusion*** occurs when a special molecule located within the plasma membrane carries a substance (usually glucose) across the membrane by means of diffusion —that is, from an area of greater concentration to one of lesser concentration
b. *Active transport* involves the movement of molecules across the plasma membrane and against a concentration gradient (from an area of lesser concentration to one of greater concentration) with the use of energy (in the form of ATP), such as by carrier-protein transport and the sodium-potassium pump
(1) In ***carrier-protein transport,*** an enzyme-like carrier protein recognizes a molecule, binds with the molecule at an active site, changes shape, then rotates through the lipid layer of the plasma membrane carrying the molecule with it
(2) The ***sodium-potassium pump*** is a special transport protein in the plasma membrane of animal cells that actively transports sodium out and potassium into the cell against a concentration gradient
c. *Exocytosis* involves the spilling of contents from a vesicle (usually from the endoplasmic reticulum or Golgi apparatus) after its membrane encounters and then fuses with the plasma membrane
d. *Endocytosis* involves the engulfment of contents by the cell after contact with the plasma membrane, which pinches inward to form a pocket
(1) *Phagocytosis* involves the engulfment of a particle into a large membrane-enclosed sac (vacuole)
(2) *Pinocytosis* involves the engulfment of droplets of extracellular fluid into a tiny vesicle

(3) *Receptor-mediated endocytosis* involves the engulfment of extracellular substances that bind with proteins embedded in the plasma membrane, which encloses to form a vesicle

E. Organelles

1. Organelles are cellular bodies suspended in cytoplasm; they have specialized functions
2. The number and function of organelles in each cell depend on the type of cell
3. ***Mitochondria,*** found only in eukaryotic cells, are responsible for cellular respiration
    a. They are composed of a smooth outer membrane and an inner membrane of folds (cristae)
    b. They take in oxygen and organic molecules and release end products (carbon dioxide and water) into the cytoplasm
    c. Mitochondria contain their own DNA
4. The *endoplasmic reticulum,* found only in eukaryotic cells, is a membranous network of tubes or channels (cisternae), which serves to transport materials throughout the cell; it can either be rough or smooth
    a. *Rough endoplasmic reticulum,* which is lined with ribosomes, helps with the production of proteins secreted by the cell (for example, in humans the rough endoplasmic reticulum of white blood cells secretes antibodies)
    b. *Smooth endoplasmic reticulum,* which has no ribosomes lining its membranes, helps with lipid synthesis, carbohydrate metabolism, and the detoxification of drugs or poisons
5. *Ribosomes,* found in all cells, are tiny granules containing ribonucleic acid (RNA) that are either distributed freely throughout the cytoplasm or attached to the endoplasmic reticulum; they are the sites where proteins and enzymes are assembled according to genetic instructions
6. The *Golgi apparatus,* found only in eukaryotic cells, is the flattened vesicle-like portion of the endoplasmic reticulum that serves as the storage area for the cell's secretory products (such as hyaluronic acid, a sticky substance that helps to bind animal cells together)
7. *Lysosomes,* found only in eukaryotic cells, are tiny membrane-bound sacs containing hydrolytic enzymes that aid in the normal decomposition of the cell and in the destruction of foreign particles
8. *Vacuoles,* found only in eukaryotic cells, are large membrane-enclosed sacs responsible for carrying out various functions
    a. Food vacuoles, which are formed by phagocytosis, fuse with lysosomes to digest food
    b. Contractile vacuoles pump excess water out of the cell
    c. Central vacuoles, found in mature plant cells, store organic compounds
9. *Plastids* are special organelles found only in plant cells
    a. Amyloplasts are colorless plastids that store starch
    b. Chromoplasts contain pigments that give fruit, flowers, and autumn leaves their color; chlorophyll, which is necessary for photosynthesis, is an example of a chromoplast

III. Cellular Organization

A. General information

1. Cells can exist independently or as aggregates in more complex structures
2. When cells aggregate, each cell tends to develop a specialized function but loses its ability to survive independently
3. Once specialized, a cell cannot return to a simpler, less specialized state
4. An *organism* is any unicellular or multicellular being capable of performing all life's basic functions and whose sum is greater than its individual parts
5. Some cells in an adult organism remain unspecialized to preserve their capacity to differentiate into a more specialized cell type when the organism must grow or undergo repair

B. Tissues

1. A tissue is an aggregate of cells in which each cell cooperates with all others in performing a group function
2. Tissues may be classified according to one of two basic types
   a. *Simple* tissue consists of only one type of cells; for example, the parenchymal tissue of plants is comprised of only parenchymal cells
   b. *Complex* tissue consists of two or more different types of cells; for example, the vascular tissue of plants is comprised of xylem and phloem

C. Organs

1. An organ is an aggregate of tissues in which each tissue cooperates in performing a group function
2. In the animal kingdom, all but the simplest animals (sponges and cnidarians) have organs

D. Organ systems

1. An organ system is an aggregate of organs in which each organ cooperates in keeping the organism alive and functioning
2. Examples of organ systems include the respiratory, circulatory, digestive, endocrine, immune, and excretory systems
3. Organ systems are found in all vertebrates and most invertebrate phyla

IV. Plant Cells

A. General information

1. As a plant grows and develops, its cells specialize to perform various functions
2. Unlike other cells, plant cells have a *cell wall,* a thick, nonliving structure that forms around the plasma membrane and encases the cell
3. Some plant cells contain both a primary cell wall (a thin, flexible structure formed while the cell is growing) and a secondary cell wall (a thick, rigid structure formed after the cell stops growing), which is located between the plasma membrane and the primary cell wall
4. The *middle lamellae,* a sticky layer located between the primary walls of adjacent plant cells, cements plant cells together

5. Plant cells may communicate with one another through cytoplasmic channels called ***plasmodesmata*** located in cell wall pits, where the secondary cell walls narrow
6. The cell wall typically plays an important part in determining the structure and function of each cell type within the plant
7. Plant cells commonly are classified according to their structure and function

B. Parenchymal cells

1. The least specialized and most abundant type of plant cells, *parenchymal cells* have the capacity to develop into other types of plant cells (under laboratory conditions, a plant can regenerate from a single parenchymal cell)
2. These cells, which are typically found packed close together, lack secondary cell walls
3. They are responsible for most of the plant's metabolism

C. Collenchymal cells

1. More specialized than parenchymal cells, *collenchymal cells* are abundant in developing leaves and stems
2. These cells lack a secondary cell wall but have a thicker primary wall than do parenchymal cells
3. They support the plant against gravity

D. Sclerenchymal cells

1. More specialized than collenchymal cells, *sclerenchymal cells* are found in areas of the plant where lengthwise growth has stopped (such as the stems and roots)
2. These cells have thick primary walls and lignin-impregnated secondary walls, which provide the plant with mechanical support (***lignin*** is the principal noncellulose component of plant tissue that provides the plant's rigidity)
3. Once mature, these cells lose their nucleus and cytoplasm and cease growing; they serve as a scaffolding on which the rest of the plant can build)
4. Sclerenchymal cells are categorized according to two types
   a. *Fibers* are elongated cells usually grouped in bundles; they provide longitudinal support throughout the plant (hemp is composed of fibers)
   b. *Sclereids,* which are shorter than fibers, are irregularly shaped (nutshells and seed coats are made up of sclereids)

E. Epidermal cells

1. Epidermal cells, which are not highly specialized, cover the external parts of the plant
2. These cells occur in diverse shapes; two distinctive types that are somewhat more specialized than the rest are guard cells and root hair cells
   a. *Guard cells,* which are sausage or crescent shaped, are found in pairs on the surface of leaves and green stems; the space between them (stoma) can be opened or closed, allowing the passage of gases into and out of the plant
   b. *Root hair cells,* which have long fingerlike extensions of cytoplasm on the side exposed to the soil, anchor the root in soil and increase the surface area available for nutrient absorption

F. Endodermal cells

1. Endodermal cells are found primarily in plant roots
2. Most of these cells contain a cell wall that is impregnated with suberin, a waterproofing material; they serve as a selective barrier that regulates the passage of substances from the soil to the roots
3. *Passage cells*—endodermal cells that do not manufacture suberin—transport water within roots or from roots to stems

G. Tracheids

1. Spindle-shaped cells containing both a primary cell wall and a secondary lignin-impregnated cell wall, tracheids are highly specialized for the transport of water
2. Comparatively large for plant cells, tracheids may grow to 5 mm long
3. Once mature, these cells lose their nucleus and cytoplasm and cease growing; they serve as a scaffolding on which the rest of the plant can build

H. Vessel elements

1. These cells are specialized for conduction of water
2. Several vessel elements can fuse together to form a single elongated column, a continuous nonliving hollow tube that can grow to several feet

I. Sieve tubes

1. Highly specialized cells, sieve tubes aid in the conduction of organic compounds
2. Each sieve tube is joined to a ***companion cell***
3. Once mature, a sieve tube loses its nucleus and cytoplasm; the companion cell then carries out all subsequent nuclear activities, such as mitosis

V. Plant Tissues

A. General information

1. Plant tissue is categorized primarily according to the type of cells contained within it
   a. *Simple* tissue consists of a single type of plant cell, such as parenchymal, collenchymal, or sclerenchymal cells, which performs a common function
   b. *Complex* tissue consists of more than one type of plant cell, such as the tracheids and vessel elements found in xylem, a highly vascular tissue that transports water and dissolved substances throughout the plant
   c. *Meristematic* tissue consists of embryonic cells that remain in an undifferentiated state as long as the plant lives
2. Plant tissue may be classified further according to where cells aggregate (such as in the epidermis, cortex, endodermis, stele, and mesophyll) and by the functions they perform

B. Epidermis

1. This tissue, which consists of epidermal cells, guard cells (in the stem) and root hair cells (in the roots), forms as a single layer of tightly packed cells on the outermost layer of the plant
2. It provides protection and nutrient absorption and secretes a waxy coating ***(cuticle)*** that helps the plant retain water

C. Cortex

1. This tissue, comprised mostly of parenchymal, collenchymal, and sclerenchymal cells, is located beneath the epidermis in roots and stems
2. It helps to transport water
3. The cortex of stems also contains chloroplasts, which enable the cells to carry out photosynthesis and produce food

D. Endodermis

1. Comprised of endodermal cells, this tissue is only one cell-layer thick
2. It separates the cortex from the central core in the roots and serves as a selective barrier regulating the passage of substances from the soil into the central core of the plant root
3. It also may transport water

E. Stele

1. The stele forms a vascular cylinder in both roots and stems and conducts water and substances throughout the plant
2. It is comprised of four types of tissues: pericycle, phloem, xylem, and pith
   a. The *pericycle,* which consists of parenchymal and sclerenchymal cells, is a potentially meristematic tissue because its cells can rapidly divide and give rise to lateral roots; it may not be present in the stem
   b. The *phloem,* which consists of parenchymal, sclerenchymal, companion, and sieve tube cells, is a vascular conducting tissue that transports organic materials over long distances in the plant
   c. The *xylem,* which consists of parenchymal cells, sclerenchymal cells, tracheids, and vessel elements, is a vascular conducting tissue that transports water and dissolved organic nutrients in the plant; the woody part of the plant, it helps support the plant against gravity
   d. The *pith,* which consists of parenchymal and sclerenchymal cells, is found in stems but not roots; it stores food and conducts water from cell to cell throughout the plant

F. Mesophyll

1. Comprised primarily of parenchymal cells, the mesophyll is equipped with chloroplasts for photosynthesis
2. Found only in leaves, mesophyll may be categorized into one of two types: palisade or spongy
   a. *Palisade* mesophyll, which is located on the upper part of the leaf, is comprised of columnar-shaped cells; because of its position on the leaf, palisade mesophyll receives the most sunlight and is a primary site for photosynthesis
   b. *Spongy* mesophyll, which is located beneath the palisade layer, contains irregularly shaped cells and numerous air spaces through which carbon dioxide and oxygen circulate

VI. Plant Organs

A. General information

1. In the plant kingdom, the differentiation of tissues into organs occurs in higher plants (such as flowering plants) only
2. Plant organs carry out specific functions, such as absorption of minerals and water, transport of water and nutrients, support against gravity, photosynthesis, and reproduction
3. The type of plant organ may depend on how tissues are arranged within the plant; for example, ***dicot*** stems contain vascular bundles (xylem and phloem) arranged in a ring between the outer cortex and the inner pith, whereas ***mono-cot*** stems contain vascular bundles arranged in a more random fashion
4. Flowers, which are found only in ***angiosperms*** (flowering seed-bearing plants), are essentially modified leaf organs

B. Root

1. The root is an aggregation of specialized tissues that allows for the movement of water, minerals, and gases into the plant by both active and passive transport
2. The root also helps move dissolved materials to other parts of the plant, anchors the plant to the soil, and stores food
3. It typically consists of an epidermis, a cortex, an endodermis, a pericycle, a phloem, and a xylem

C. Stem

1. The stem provides support for the growing plant
2. It allows for transport of materials between roots and leaves by means of the xylem, which conducts water upward to the leaves, and the phloem, which conducts soluble food materials from the leaves downward into the stems and roots
3. It also serves as a food storage area
4. The tissues constituting a stem include the epidermis, cortex, pith, xylem, and phloem
5. The stem attaches to the leaf at a juncture called a *node;* the part of the stem located between nodes is called an *internode*

D. Leaves

1. The major photosynthetic organs of most plants, leaves exist to make the food required for the plant's survival
2. Leaves have openings or gaps (stomas) between guard cells, which allow for the exchange of carbon dioxide and oxygen and for the evaporation of water (transpiration)
3. The junction at which a leaf attaches to a stem at a node is called a *petiole*
4. The tissues found in a typical leaf include the epidermis, mesophyll, xylem, and phloem

VII. Animal Tissues

A. General information

1. Animal tissues are found only in "true" animals—that is, all animals belonging to the kingdom *Animalia;* some organisms with animal pattern cells, such as those belonging to the kingdom *Protista* (such as amoeba, ciliates, and flagellates) are unicellular and therefore have no tissues
2. Animal tissue is composed of similar cells that are bound together by a sticky coating or woven together into a fabric of extracellular fibers
3. The animal tissues found in ***vertebrates*** (the class to which humans belong) can be classified into four major categories: epithelial, connective, muscle, and nerve

B. Epithelial tissue

1. Epithelial tissue is comprised of epithelial cells that are cemented directly to one another by an organic compound whose main component is hyaluronic acid
2. This tissue forms the covering or lining of body surfaces and provides protection against mechanical injury, invading microorganisms, and fluid loss; it also lubricates the skin and facilitates chemical absorption
3. Some epithelial tissue is highly specialized to secrete enzymes and hormones (for example, glandular epithelium) or to receive sensations, such as smell and sight (for example, sensory epithelium)
4. Epithelial tissue may be classified as simple or stratified, depending on the layers of cells
   a. *Simple* epithelium is comprised of a single layer of squamous, cuboidal, or columnar cells
      (1) *Simple squamous* epithelium is specialized tissue that regulates the exchange of materials between tissues; it is found on the inner lining of blood vessels (endothelium) and lung air sacs, where it facilitates the diffusion of oxygen and carbon dioxide between blood and surrounding tissue; some simple squamous epithelium also is found along the lining of the mouth and esophagus, where it facilitates the secretion of mucus—a substance that helps to lubricate these passages and protect them from injury and infection
      (2) *Simple cuboidal* epithelium aids in the secretion of physiologically active substances, such as hormones; it is found in the epithelia of kidney tubules and many glands, including the thyroid and salivary glands
      (3) *Simple columnar* epithelium typically is located where secretion (particularly of mucus) or active absorption of substances is required, such as along the lining of the stomach and intestines; a special type of columnar epithelium that contains cilia (*ciliated columnar* epithelium) is found in the respiratory tract, where it is especially adapted to sweeping out bacteria and dust
   b. *Stratified* epithelium, which is comprised of multiple layers of squamous, cuboidal, or columnar cells, is typically found on surfaces subject to abrasion (such as the skin and along the lining of the espohagus, anus, and vagina); *stratified squamous* epithelium, which constitutes the epidermis of the skin, is the most common type

C. Connective tissue

1. Connective tissue is comprised of widely dispersed cells embedded in an extensive intercellular network of fibrous material and matrix
   a. The cells found in connective tissue are highly specialized to perform specific functions
      (1) *Fibroblasts* secrete the intracellular components of the tissue (fibers and matrix)
      (2) *Macrophages* are irregularly shaped cells that are capable of ameboid locomotion and the engulfment of foreign material
      (3) *Fat cells* are capable of storing a large droplet of fat, which occupies most of the cell space
      (4) *Mesenchymal cells,* embryonic, unspecialized cells that are capable of differentiating into connective tissue cells, play an important role in healing and regeneration
      (5) *Chondrocytes,* which are found in cartilage, secrete a firm, gel-like substance called *chondrin*
      (6) *Osteocytes,* which are found in bone, release calcium phosphate
      (7) *White blood cells, red blood cells,* and *platelets* are found in blood; they respectively help fight infection; transport oxygen, carbon dioxide, and nutrients; and form clots
   b. The fibrous material found in the intercellular portion of connective tissue may be classified as one of three types: collagenous, elastic, or reticular
      (1) *Collagenous fibers* (also called white fibers) are composed of the protein collagen; they are flexible but resist stretching and therefore help to strengthen tissue
      (2) *Elastic fibers* (also called yellow fibers) are composed of the protein elastin; they can be easily stretched and consequently help give the tissue a certain resilience
      (3) *Reticular fibers* branch and interlock to form complex networks that help join connective tissue with adjacent tissues
   c. The matrix found in the intercellular portion of connective tissue is an amorphous, nonliving substance consisting of water, protein, carbohydrates, and lipids; it can vary in form —from liquid to semisolid to solid
2. Connective tissue may be classified according to one of six major types
   a. *Loose connective tissue* consists of fibroblasts, macrophages, fibrous material (collagen, elastin, or reticular fibers), and a soft, jelly-like matrix; found immediately beneath most epithelial membranes, this type of tissue provides flexibility and protects against organism infiltration
   b. *Adipose tissue* consists of fat cells closely packed with fibrous material (either collagen or elastin); found under the skin and around the kidneys and eyeballs, this tissue serves as a reserve food supply, insulates against heat, and supports and protects various organs
   c. *Dense connective tissue* is made up of fibroblasts with fibrous material (mostly collagen with some elastin); found in tendons and ligaments, this type of tissue resists pulling and stretching and helps strengthen surrounding tissue
   d. *Cartilage* consists of chondrocytes embedded in a firm gel-like substance (chondrin); it covers the articulating ends of bones, allowing for frictionless

movement, and serves as a shock absorber (such as in intervertebral disks)

e. *Bone* is composed of osteocytes, collagen, and a matrix of calcium phosphate; it supports the body and protects organs

f. *Blood* consists of red and white blood cells, platelets, and an intercellular material composed of plasma that contains water, dissolved proteins, salts, nutrients, antibodies, hormones, and wastes; it transports oxygen and nutrients to other cells of the organism, transports wastes away from cells, defends the organism against foreign cells, and forms clots

D. Muscle tissue

1. Muscle tissue is capable of contraction and movement
2. It is made up of muscle cells comprised of large numbers of microfilaments containing the contractile proteins actin and myosin
3. Muscle tissue can be classified according to one of three types: voluntary, smooth, or cardiac
   a. *Voluntary* (striated) muscle contracts according to the organism's will
   b. *Smooth* muscle contracts involuntarily
   c. *Cardiac* muscle, which is found only in the heart, contracts involuntarily

E. Nerve tissue

1. Nerve tissue is specialized to transmit messages or impulses throughout the body
2. It consists of specialized cells called *neurons* that contain numerous tiny projections ***(dendrites),*** which conduct impulses toward the cell body, and a single long extension ***(axon),*** which transmits impulses away from the cell body

Study Activities

1. Select five basic cell functions and identify the organelles that perform those functions.
2. Draw a basic plant and animal cell, labeling all structures and organelles. Note the major differences between the two cell types.
3. Compare passive and active transport, giving two examples of each.
4. Outline the relationship among cells, tissues, organs, and organ systems.
5. Describe the structure and function of the eight types of plant cells.
6. Identify the five types of animal tissue and give an example of each from the human body.

4

Reproduction and Growth

Objectives

After studying this chapter, the reader should be able to:
- Write a short essay on the differences between mitosis and meiosis.
- Identify four types of asexual reproduction.
- Outline what is meant by alternation of generations.
- Describe the role and types of meristematic tissue.
- Discuss the stages of early embryonic development in animals.
- List the three types of chromosomal alterations and discuss their consequences.

I. The Beginning of Life

A. General information

1. The ability to reproduce distinguishes living things from nonliving things
2. All reproduction involves cell division; a cell first must copy all of its *genes* (discrete units of hereditary information comprised of deoxyribonucleic acid [DNA]) before it divides, then allocate this genetic material equally between each daughter cell
3. The two major types of reproduction are asexual and sexual
    a. In *asexual reproduction,* only one parent is involved; the resulting cell is an exact copy of its parent
    b. In *sexual reproduction,* two parents are necessary; the resulting cell is a fusion of two parent cells and is not exactly identical to either parent
4. In prokaryotic cells (those without a nucleus, such as bacteria), cell division involves the duplication of DNA, followed by cytokinesis (division of the cytoplasm); this process is called *binary fission*
5. In eukaryotic cells (those with a nucleus), cell division involves the division of the nucleus, followed by cytokinesis; this process is called *mitosis*
6. Cell division allows unicellular organisms to duplicate themselves entirely and allows multicellular organisms to grow, develop from a single cell into a multicellular adult organism, and repair and renew cells

B. Chromosomes

1. *Chromosomes* —long, threadlike aggregations of genes —are located in the nucleus of every eukaryotic cell (in bacteria, most of the DNA is found in a single circular molecule called the *bacterial chromosome;* it is a simple structure and

contains fewer proteins than a eukaryotic chromosome; many bacteria also have plasmids —small rings of DNA that carry accessory genes)

a. Chromosomes occur naturally in pairs

b. The chromosomes in each pair are identical (homologous) to each other in several aspects —length, staining pattern, and type of genes; such chromosome pairs are called *autosomes*

c. In organisms with separate sexes, the two chromosomes that carry the genes for sex determination (called the *sex chromosomes*) are not a homologous pair but remain together as a unit

d. In humans, each cell contains 23 pairs of chromosomes, 22 of which are homologous and one that is not (the sex chromosomes, designated X and Y)

2. When chromosomes replicate, each gene also replicates
3. In asexual reproduction, the organism's characteristics do not vary from generation to generation; in sexual reproduction, the genetic material from the two parents may be shuffled, resulting in varied characteristics from one generation to the next

C. Mechanism of sexual reproduction

1. Sexual reproduction requires a mechanism for reducing the number of chromosomes in half (*meiosis*) and a mechanism for fusing two cells into one (***fertilization***)
2. The cells that fuse together, called ***gametes,*** contain half the number of chromosomes
   a. Each gamete contains one ***homologous chromosome*** and either an X or a Y sex chromosome
   b. In humans, gametes contain 23 unpaired chromosomes
   c. A *haploid* cell contains half the number of chromosomes; a *diploid* cell contains a full number of chromosomes (in humans, the haploid number [designated as "n"] is 23, the diploid number [designated as "2n"] is 46)
3. Fertilization occurs when a female gamete (ovum) fuses with a male gamete (spermatozoon) to form a new organism
4. The fertilized ovum, or *zygote,* contains the diploid number of chromosomes

D. Patterns of sexual reproduction

1. Sexual life cycles in living organisms vary according to the timing of life cycle events (meiosis and fertilization) and whether the adult organism is haploid or diploid
2. Three major variations are seen in sexual life cycles
   a. In one pattern (typical of most animals, including humans), adults are multicellular diploid organisms; meiosis produces haploid gametes, gametes fuse to produce a diploid zygote, and mitotic divisions lead to a multicellular adult
   b. In another pattern (typical of many fungi and some *Protista*), adults are haploid organisms; mitosis produces haploid gametes, gametes fuse to produce a diploid zygote, and meiosis immediately occurs to produce haploid cells that then grow by mitosis into a multicellular adult
   c. In a different pattern (typical of all plants and some algae), both asexual and sexual reproduction is combined in such a way that one adult generation is a multicellular diploid organism (called a sporophyte) and the next gener-

ation is a multicellular haploid organism (called the ***gametophyte*** generation); this pattern is called *alternation of generations*

II. Mitosis

A. General information

1. Mitosis is the cellular mechanism by which the nuclear content of eukaryotic cells reproduces and divides
2. It results in the formation of two daughter cells from one parent cell
3. Mitotic cell division occurs in five phases: interphase, prophase, metaphase, anaphase, and telophase (see *Mitosis* for an illustration of these steps)
4. The sequence of ordered events that occurs between the time a cell divides to form two daughter cells and the time those cells divide again is called the *cell cycle;* this sequence encompasses all of the phases in mitosis

B. Interphase

1. In *interphase,* commonly called the resting stage, the nondividing cell appears to be resting but actually undergoes normal cellular functions; during this phase, which constitutes about 90% of the cell cycle, genetic material is replicated in preparation for the next division sequence
2. Interphase can be divided into four subphases based on the timing of DNA synthesis
   a. The G1 phase (also called the presynthesis gap) marks the time when RNA synthesis, protein synthesis, and cell growth occur in preparation for DNA synthesis
   b. The G0 phase (or dormant stage) is when all the cell's activities occur except for reproduction and growth; not all cells enter into this phase, and the amount of time spent in the phase varies depending on the cell type
   c. The S phase (or synthesis phase) is the time when DNA synthesis occurs; the amount of DNA in the cell's nucleus doubles
   d. The G2 phase (also called the postsynthesis gap) is when DNA synthesis ceases; RNA synthesis and protein synthesis occur in preparation for mitotic division
3. The nucleus and nuclear membrane are well defined during this phase, and the nucleolus is typically prominent
4. The chromosomes —which are long and thin but appear under a microscope as only a mass of granular *chromatin* (an indistinguishable mix of proteins and DNA within the nucleus) —replicate by splitting lengthwise but remain attached at the center
5. In animal cells, two pairs of *centrioles* —cylindrically shaped cytoplasmic organelles — appear just outside the nucleus at this time

C. Prophase

1. *Prophase* marks the beginning of the mitotic phase of the cell cycle, when actual cell division occurs
2. During this phase, the nucleolus disappears and the chromatin shortens and thickens into distinct chromosomes

Mitosis

In mitosis, the nuclear contents of a cell reproduce and divide, resulting in the formation of two new daughter cells, each containing the diploid number of chromosomes. The five steps, or phases, are illustrated below.

Interphase

- The nucleus and nuclear membrane are well defined; the nucleolus is prominent
- Chromosomes replicate, each forming a double strand that remains attached at the center by a centromere; they appear as only an indistinguishable matrix within the nucleus
- Centrioles (in animal cells only) appear outside the nucleus

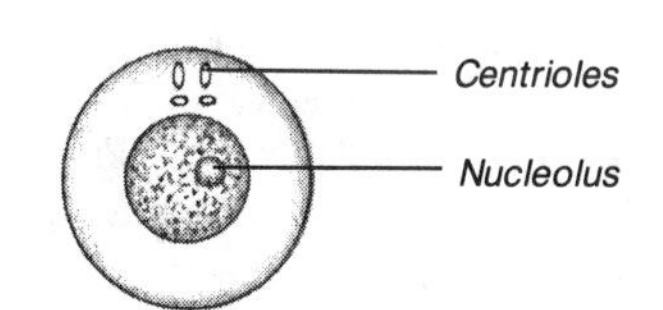

Prophase

- The nucleolus disappears
- Chromosomes become distinct; the halves of each duplicated chromosome (chromatids) remain attached by a centromere
- Centrioles move to opposite sides of the cell and radiate spindle fibers (in plant cells, spindle fibers radiate from concentrated areas within the cytoplasm)

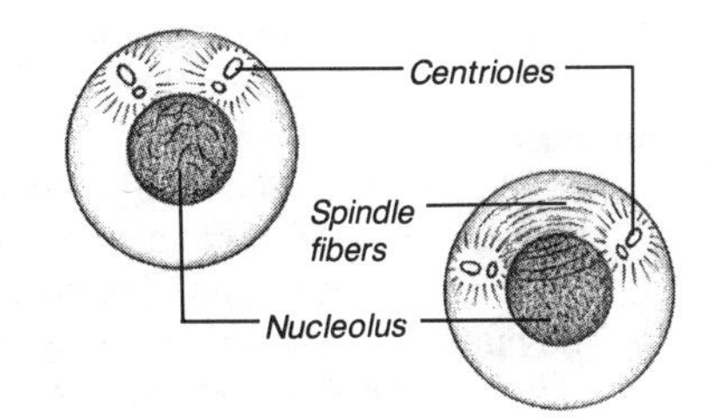

Metaphase

- Chromosomes line up randomly in the center of the cell between the spindles, along the metaphase plate
- The centromere of each chromosome replicates

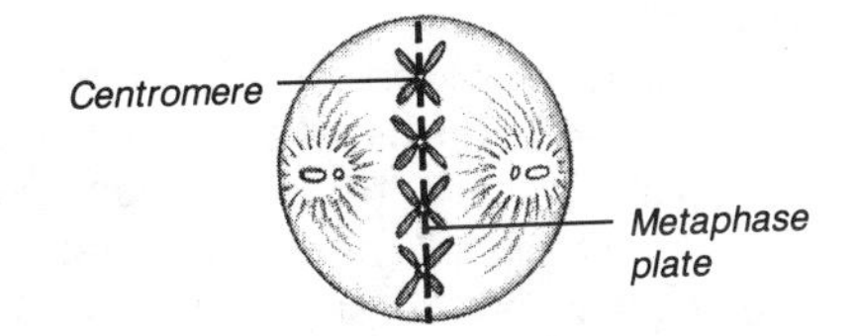

Anaphase

- Centromeres move apart, pulling the separated chromatids (now called chromosomes) to opposite ends of the cell
- The number of chromosomes at each end of the cell equals the original number

Telophase

- A nuclear membrane forms around each end of the cell; spindle fibers disappear
- The cytoplasm compresses and divides the cell in half; each new cell contains the diploid number of chromosomes

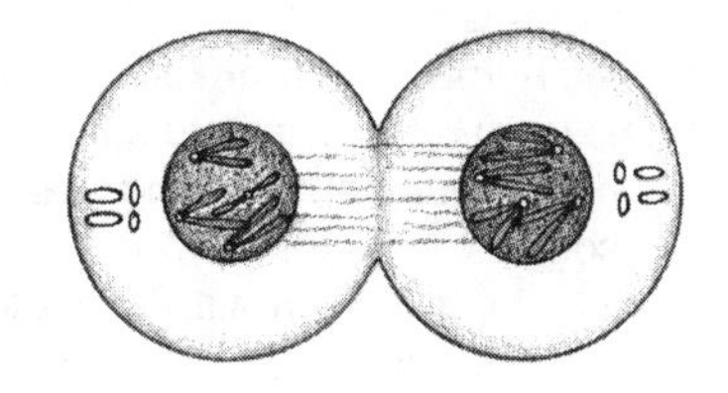

3. As the nuclear membrane disappears, the chromosomes —which replicated during interphase but still remain attached at the center by a ***centromere***—appear randomly distributed throughout the cytoplasm; each half of a replicated chromosome is called a *chromatid*
4. In animal cells, the centrioles move separately toward opposite ends of the cell and begin radiating fine threads of cytoplasm *(spindle fibers)* that attach to the centromere of each chromosome; each radiating mass of spindle fibers and centriole is called a *mitotic spindle*
5. In plant cells (which do not contain centrioles), spindle fibers radiate from concentrated parts of the cytoplasm that are functionally equivalent to the centrioles of animal cells

D. Metaphase

1. During *metaphase,* the chromosomes line up in the center of the cell, equidistant between the spindle's two poles, on the *metaphase plate*
2. Then, the centromere of each chromosome replicates, allowing the chromatids to separate from one another

E. Anaphase

1. During *anaphase,* the centromeres move apart, pulling one chromatid from each chromosome to the opposite end of the cell
2. At this point, each separated chromatid is considered a complete chromosome
3. The number of chromosomes at each end of the cell is identical to the original number of cell chromosomes

F. Telophase

1. During *telophase,* a nuclear membrane forms around the chromosomes collected at each end of the cell, and the spindle fibers disappear
2. In animal cells, the cell membrane compresses in the middle of the cell, forming a cytoplasmic *cleavage furrow* that divides the cell into two new daughter cells
3. In plant cells, a cell plate forms in the middle of the cell, and particles of cellulose are deposited to form a new cell wall

III. Meiosis

A. General information

1. Meiosis is the cellular mechanism that reduces by half the number of chromosomes of sex cells and enables sexual reproduction to occur
2. Meiosis is necessary for all sexually reproducing organisms because it prevents the number of chromosomes from doubling from one generation to the next
3. The union of two haploid sex cells (*fertilization*) restores the chromosomes to the diploid number, ensuring that the offspring has one complete set of chromosomes
4. In plants, meiosis results in the formation of haploid spores; in animals, the formation of gametes
5. *Gametogenesis* is the meiotic process by which a diploid cell reduces its number of chromosomes to produce a haploid gamete

(text continues on page 37)

Meiosis

In meiosis, which occurs only in sex cells, the number of chromosomes are reduced by half. Meiosis involves two division sequences separated by a resting phase and results in the formation of four new cells, each having the haploid number of chromosomes.

FIRST DIVISION

This stage begins with one parent cell and ends with two daughter cells, each containing the haploid number of chromosomes.

Interphase

- Chromosomes replicate, each forming a double strand that remains attached at the center by a centromere; the chromosomes appear as only an indistinguishable matrix within the nucleus
- Centrioles (in animal cells only) appear outside the nucleus

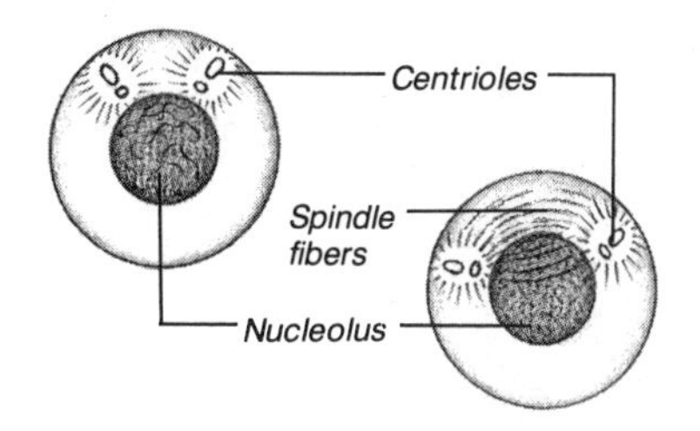

Prophase I

- The nucleolus and nuclear membrane disappear
- Chromosomes become distinct; chromatids remain attached by a centromere
- Homologous chromosomes move close together and interwine—a process called synapsis; exchange of genetic information (genetic recombination) may occur
- Centrioles separate, and spindle fibers appear

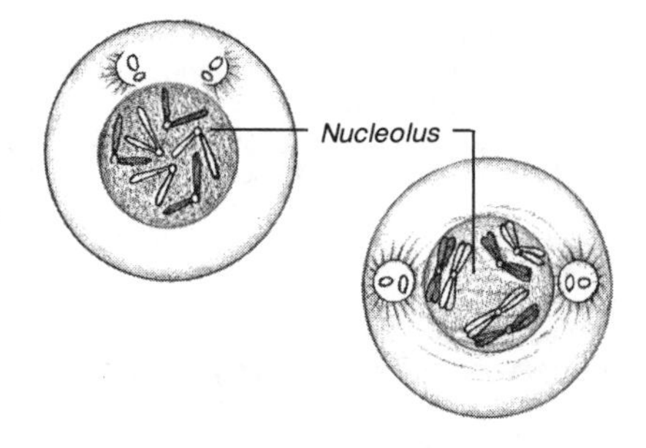

Metaphase I

- Pairs of synaptic chromosomes line up randomly along the metaphase plate
- Spindle fibers attach to each chromosome pair; centromeres do not replicate

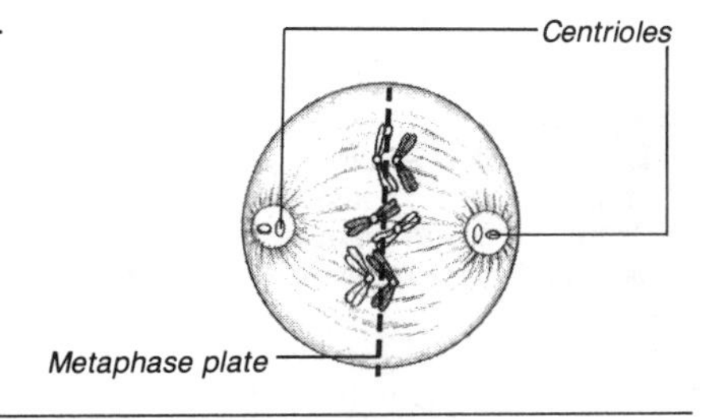

Anaphase I

- The synaptic pairs separate, with spindle fibers pulling the homologous, double-stranded chromosomes to opposite ends of the cell
- Because the centromeres have not divided, the chromatids remain attached

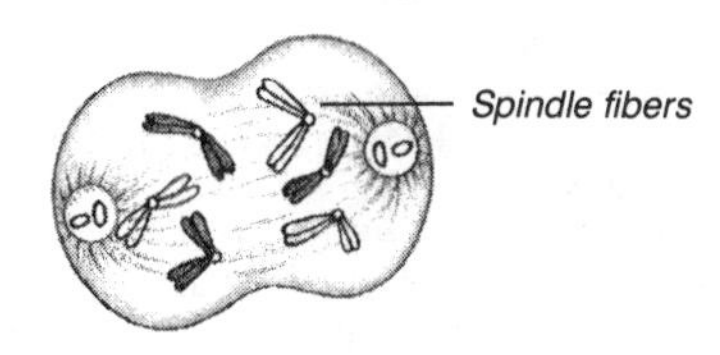

(continued)

Meiosis *(continued)*

Telophase I

- A nuclear membrane forms around each end of the cell
- The spindle fibers and chromosomes disappear
- The cytoplasm compresses and divides the cell in half; each new cell contains the haploid number of chromosomes

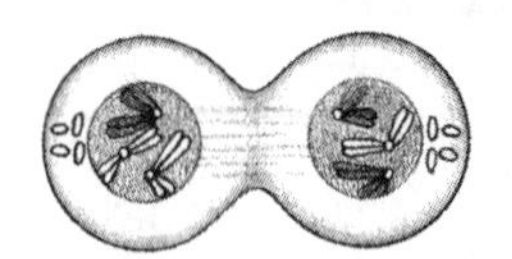

Interkinesis

- The nucleus and nuclear membrane are well defined; the nucleolus is prominent
- Each chromosome consists of two chromatids that do not replicate

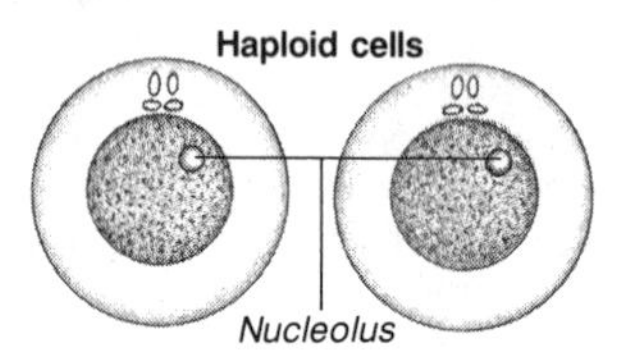

SECOND DIVISION

This stage is essentially a mitotic division. It begins with two new daughter cells, each containing the haploid number of chromosomes, and ends with four new haploid cells.

Prophase II

- The nuclear membrane disappears, and spindle fibers form
- Double-stranded chromosomes appear as thin threads; they do not synapse

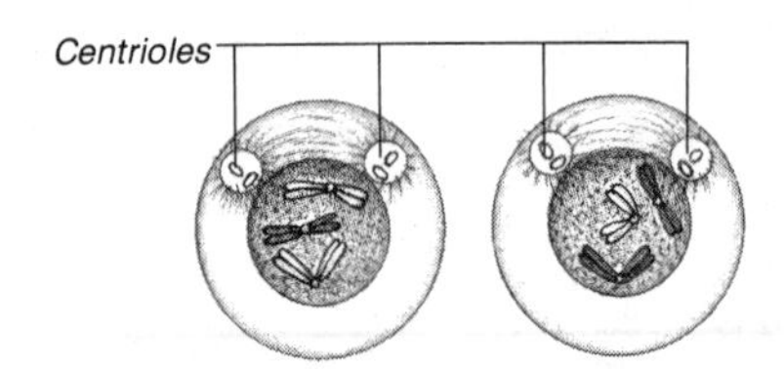

Metaphase II

- Chromosomes line up along the metaphase plate
- Centromeres replicate

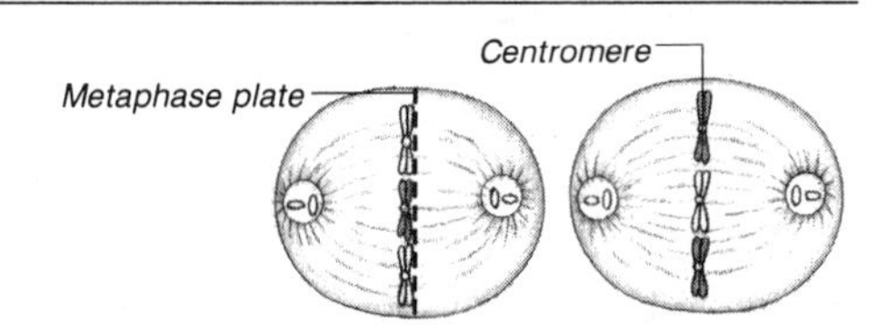

Anaphase II

- The chromatids separate; each is now a single-stranded chromosome
- Chromosomes move away from one another and pull to opposite ends of the cell

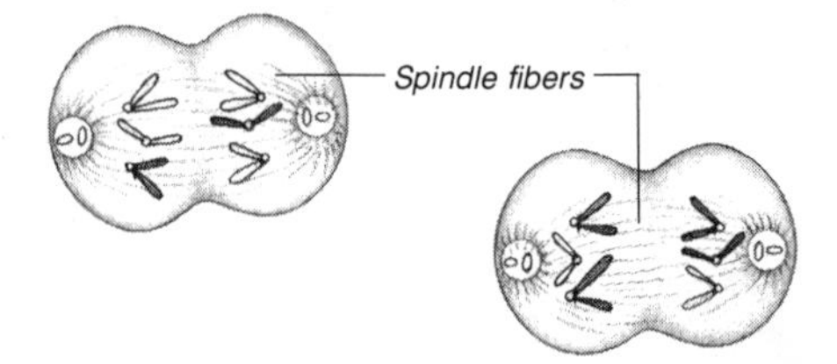

Telophase II

- A nuclear membrane forms around the end of each cell; the chromosomes and spindle fibers disappear
- The cytoplasm of each cell compresses and divides the cells in half, leaving four new daughter cells each containing the haploid number of chromosomes

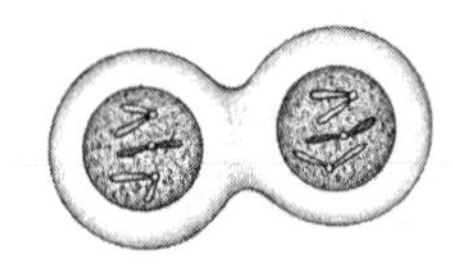
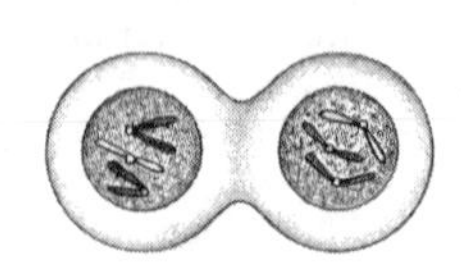

a. *Spermatogenesis* describes the process that results in a male gamete (spermatozoon)
b. *Oogenesis* describes the process that results in a female gamete (ovum)

6. Meiosis involves two successive division sequences that result in four new cells, each of which is haploid (see *Meiosis* for an illustration of these sequences)
   a. The first division reduces the number of chromosomes by half and ends with the formation of two haploid cells
   b. The second division is essentially a mitotic division of these two haploid cells

B. First meiotic division

1. This first division, which encompasses all of the phases as in mitosis, results in the formation of two daughter cells, each containing the haploid number of chromosomes
2. *Interphase* is similar to that in mitotic division
   a. The nucleus and nuclear membrane are well defined, and the nucleolus is typically prominent
   b. Chromosomes, which remain indistinguishable as chromatin, replicate, resulting in two sets of double-stranded chromosomes
   c. In animal cells, two pairs of centrioles appear just outside the nucleus
3. During *prophase I,* the chromosomes gradually appear as distinct structures, and the nucleolus and nuclear membrane disappear; the centrioles separate, and spindle fibers appear; also, *synapsis*—a phenomenon that occurs only in meiosis, not mitosis—takes place
   a. During synapsis, the members of each pair of homologous chromosomes move close together; they intertwine but do not fuse
   b. While the chromosomes are intertwined, genetic information is exchanged between the paired homologous chromosomes
   c. The crossing over of genetic information results in gene combinations different from those of previous generations; this exchange, called *genetic recombination,* greatly increases the genetic variability of the offspring
4. During *metaphase I,* each synaptic pair of chromosomes lines up randomly along the metaphase plate; spindle fibers attach to each chromosome pair (each pair consists of two joined chromatids that do not separate); the centromeres do not replicate
5. During *anaphase I,* the synaptic pairs split, pulling each set of double-stranded chromosomes to opposite ends of the cell; because the centromeres have not divided, the chromatids remain attached
6. During *telophase I,* the double-stranded chromosomes disappear and two new nuclei form; the cells begin to separate

C. Interkinesis

1. *Interkinesis* is the resting stage that occurs between the first and second meiotic divisions
2. Although similar to interphase, interkinesis does not involve the replication of genetic material; each chromosome consists of two chromatids that do not replicate

D. Second meiotic division

1. During this stage, which consists of the same phases as in mitotic division, the two daughter cells divide to form four new cells, each containing the haploid number of chromosomes
2. During *prophase II,* the double-stranded chromosomes appear as thin threads, the nuclear membrane disappears, and the spindle forms; the chromosomes do not synapse because the nucleus is haploid and no homologous chromosomes are present
3. During *metaphase II,* the chromosomes line up along the metaphase plate and the centromeres replicate
4. During *anaphase II,* the single-stranded chromatids move away from each other and pull to opposite ends of the cell
5. During *telophase II,* membranes begin to form around the chromosomes, the spindles disappear, and the cell divides into individual cells, each containing the haploid number of chromosomes

IV. Asexual Reproduction

A. General information

1. Asexual reproduction requires only one parent, which passes on all of its genes to its offspring
2. No genetic variation is possible in asexual reproduction
3. Types of asexual reproduction include binary fission, simple mitotic division, budding, fragmentation, regeneration, and parthenogenesis

B. Binary fission

1. Found in such prokaryotes as bacteria and blue-green algae, this type of reproduction begins when a single cell splits into two daughter cells
2. Cytokinesis results in an even distribution of cytoplasm between the two daughter cells
3. Because prokaryotic cells have no nuclei, mitosis (which involves nuclear division) does not occur
4. Each new cell is independent, grows to its maximum size, then divides again

C. Simple mitotic division

1. Many unicellular eukaryotic organisms, such as those in the *Protista* kingdom, undergo a simple mitotic division to produce two identical daughter cells
2. This type of division encompasses all of the phases of mitosis, including interphase, prophase, metaphase, anaphase, and telophase (see *Mitosis,* page 33)

D. Budding

1. Budding, which occurs in single-celled organisms (such as yeast) and simple multicellular organisms (such as hydra), results in the formation and growth of a new organism directly out of the parent's body
2. In yeast, the bud remains attached; in hydra, the bud eventually separates from the parent

E. Fragmentation

1. This reproductive method involves the breaking of the parent's body into several pieces, each of which develops into a complete adult
2. Certain types of segmented worms reproduce by fragmentation

F. Regeneration

1. Regeneration is the ability to grow back a missing part or to produce an entire organism from a part after injury
2. Sponges have extensive regenerative abilities, which they use not only for the repair of lost parts but also to reproduce asexually from fragments broken off from a parent sponge
3. The starfish also has impressive regenerative capabilities; if a starfish is cut into several pieces, a new organism will develop from each piece that has a part of the central disk attached

G. Parthenogenesis

1. Parthenogenesis results in the development of an egg without fertilization
2. Adults produced by parthenogenesis are haploid and do not undergo meiosis to form new eggs
3. This pattern of reproduction occurs in aphids, rotifers, and daphnia at certain times in their life cycle; at other times, they reproduce sexually

V. Alternation of Generations

A. General information

1. This type of reproduction involves a life cycle in which the generations alternate between haploid (gametophyte generation) and diploid (sporophyte generation) states
2. It involves both asexual and sexual reproduction at different stages in the life cycle
3. Plants, fungi, and some multicellular algae reproduce through alternation of generations
4. Ulva, a multicellular algae, undergoes alternation of generations; the adult gametophyte and the adult sporophyte are *isomorphic*—similar in shape and structure
5. In most plants, the adult gametophyte and the adult sporophyte are *heteromorphic*—different in shape and structure

B. Reproductive stages

1. Meiosis occurs in the sporophyte generation to produce haploid cells called ***spores***, which differ from gametes in that they do not fuse with another spore but divide mitotically to produce a mature adult
2. The mature adult produced from the spore is a multicellular haploid organism (the gametophyte generation)
3. The gametophytes produce gametes through mitosis; these gametes fuse to produce a diploid zygote
4. The diploid zygote grows into a multicellular diploid adult (***sporophyte***)

VI. Sexual Reproduction in Lower Organisms

A. General information

1. Sexual reproduction in lower organisms involves either the temporary exchange of genetic information or the production of both eggs and sperm by one individual
2. Conjugation and hermaphroditism are the primary methods of reproduction among lower organisms

B. Conjugation

1. *Conjugation* involves the exchange of genetic material between two morphologically indistinguishable cells that are temporarily joined; the fusion of such cells is called ***isogamy***
2. This form of reproduction produces gametes that are similar (in contrast, fertilization produces gametes that are distinguished as male or female)
3. Spirogyra, a type of green algae, are an example of an organism that reproduces through conjugation
   a. Two filaments —cells arranged lengthwise in a long thread —lie side by side
   b. Protuberances develop on the sides of the cells that are in contact with each other
   c. The walls between the protuberances of each cell disintegrate, forming a conjugation tube
   d. One haploid cell becomes ameboid (capable of moving by means of elongation and retraction of cellular extensions called *pseudopods*), moves through the conjugation tube, and fuses with another haploid cell, thereby forming a zygote

C. Hermaphroditism

1. A hermaphroditic organism produces both ova and spermatozoa
2. Some hermaphrodites fertilize themselves, whereas others must mate with another member of the same species; in these cases, each organism serves as both male and female, donating and receiving sperm
3. Hermaphroditism is common among sessile or burrowing animals (such as barnacles and earthworms) and parasites (such as tapeworms)
4. In *sequential hermaphroditism*, a variation of hermaphroditism, an individual reverses its sex during its lifetime; this pattern is found in certain species of reef fish

VII. Reproduction in Higher Plants

A. General information

1. The higher plants, or *angiosperms,* depend on specialized organs —flowers — for reproduction
2. Angiosperms are classified into families largely based on the structure and function of their flowers (see Appendix B: *Taxonomic Classification of Living Organisms*)
3. The flower, which develops on the adult sporophyte, produces microspores and megaspores

a. *Microspores* develop into male gametophytes; immature male gametophytes are called *pollen grains*
b. *Megaspores* develop into female gametophytes, which consist of an embryo sac that contains the ovum

4. Any organism that produces two different types of spores is considered *heterosporous*
5. ***Pollination*** is the mechanism by which higher plants transfer spores
6. The successful transfer of a male spore (pollen grain) to a female spore-containing site (stigma) results in fertilization
7. After fertilization, the ovule develops into a ***seed;*** the ovary, into a fruit to nourish the seed

B. Reproductive organs of a flower

1. The flower—the reproductive structure of an angiosperm—is comprised of several parts, not all of which are directly involved in reproduction (see *Reproductive Organs of a Flowering Plant* for an illustration)
2. The sterile (nonreproductive) parts are the petals and sepals

Reproductive Organs of a Flowering Plant

This illustration shows a cross-section of the major parts of a flower, which comprises the reproductive organs of a flowering plant (angiosperm).

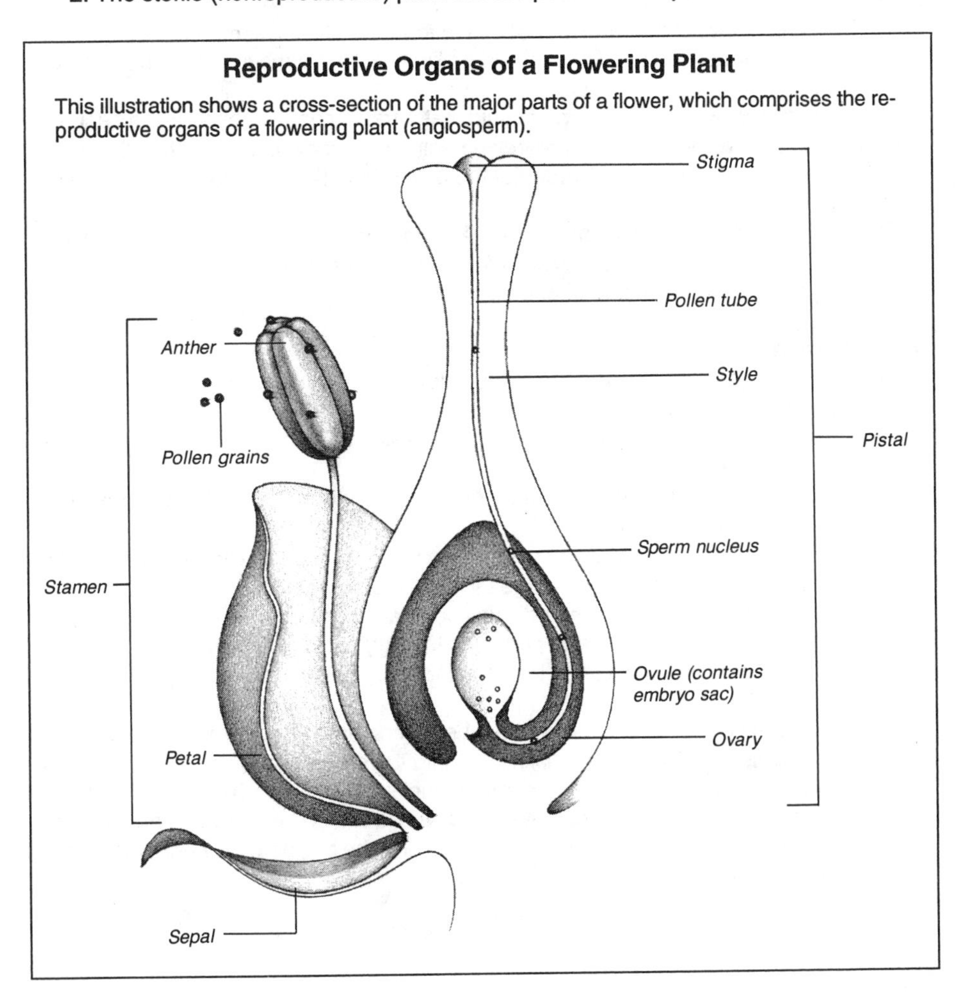

a. *Petals* are the brightly colored parts of the flower; a group of petals in a circular arrangement is called a *corolla*
b. *Sepals* are small, green, leaflike structures that form an outer layer at the base of the petals; they collectively make up the *calyx* (at one time they enclosed the flower bud)

3. The main reproductive parts of a flower are the stamen and the pistil
   a. The *stamen* consists of a slender stalk (the filament) and an enlarged top (the anther or pollen case); the anther contains the male gametophyte, or pollen grain
   b. The *pistil* (also called a *carpel*) is located within the center of the flower
      (1) It consists of an enlarged portion at the bottom (the ovary), a slender portion that extends up (the style), and a sticky portion at the top of the style (the stigma)
      (2) The ovary encloses one or more small structures called *ovules,* which contain the female gametophyte, or embryo sac

C. Pollination

1. Pollination refers to the transfer of pollen from the plant's stamen to its stigma; it can occur as a result of wind (wind pollination) or the movement of birds and insects
2. Flowers secrete nectar (a sweet liquid that attracts hummingbirds and insects) from the flower base; nectar-seeking animals become coated with pollen grains, then inadvertently transfer the pollen from flower to flower through their movements
3. A flower can be pollinated through *self-pollination* (the transfer of pollen within the same flower) or *cross-pollination* (the transfer of pollen from the stamen of one flower to the stigma of another)

D. Fertilization

1. When deposited on the stigma, each pollen grain (a single cell with two nuclei) begins to germinate
2. On germination, each pollen grain forms a pollen tube that begins to grow down through the style
   a. Of the two nuclei, one (the tube nucleus) disintegrates shortly after the pollen tube forms
   b. The other nucleus divides to form two nuclei called the *sperm nuclei*
3. Within each ovule, one cell enlarges to form the *embryo sac*
   a. The embryo sac's nucleus divides several times to form eight nuclei
   b. Of these eight nuclei, one becomes the egg nucleus and two unite to form an endosperm nucleus; the remaining five nuclei have no further significant role
4. Upon reaching the ovule, the pollen tube enters through a tiny opening called the *micropyle,* and the sperm nuclei pass into the embryo sac
5. Double fertilization occurs—one sperm nucleus fuses with the egg nucleus to form a zygote; the other sperm nucleus fuses with the endosperm nucleus to form the endosperm that will provide food for the zygote (see *Fertilization in Higher Plants*)

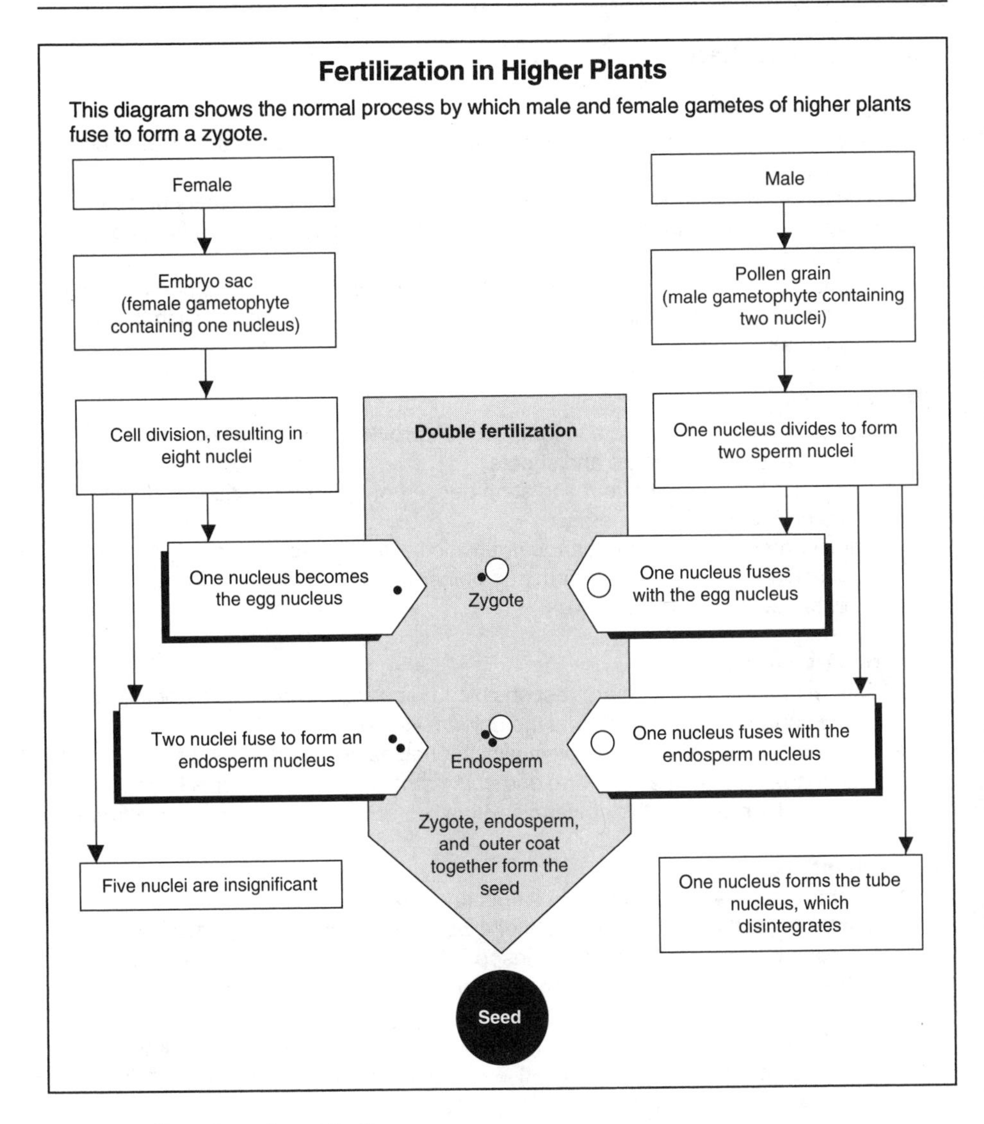

E. Formation of seed and fruit

1. After fertilization occurs, the ovary eventually develops into a fruit, and the ovule becomes a seed
2. As the seed develops, the plant embryo begins to differentiate into distinct organs
3. Two cotyledons (embryonic leaves) form to store food
   a. The epicotyl becomes the upper part of the stem and the leaves
   b. The hypocotyl develops into the lower part of the stem and the roots
4. After a seed has formed, the embryo usually becomes dormant until conditions are right for germination

VIII. Growth in Plants

A. General information

1. Special embryonic tissue, called *meristematic tissue,* makes plant growth possible
2. The three types of meristematic tissue are apical meristems, primary meristems, and lateral meristems; both primary and lateral meristems derive from apical meristems
3. Plants grow in two ways —*primary growth,* in which the plant lengthens and cells differentiate into specialized tissues, and *secondary growth,* in which the plant increases in girth

B. Primary growth

1. The cells needed for a plant to lengthen are provided by the apical meristems, located at the tips of roots and shoots
2. The major tissues of the root and stem derive from primary meristems, located behind the apical meristems
3. Primary meristems include *protoderm* (which gives rise to epidermis), *procambium* (which gives rise to xylem and phloem), and *ground meristem* (which gives rise to pith and cortex)

C. Secondary growth

1. Lateral meristems, which are responsible for secondary growth, are located in roots and stems and surround the phloem tissue
2. They consist of *vascular cambium* (which enables lateral growth by giving rise to secondary phloem and xylem) and *cork cambium* (which enables lateral growth by replacing the epidermis with a protective coat called the *periderm*)

D. Growth of shoots

1. Shoot systems are comprised of stems and leaves
2. Inspection of the external structure of a woody plant stem during the winter —a time of relative dormancy when inspection is easy because the plant has lost its leaves —reveals the presence of terminal buds, axillary buds, leaf scars, and terminal bud scale scars
   a. *Terminal buds* are situated at the ends of shoots; in spring, they shed their scales (protective coverings) and begin growing by producing a series of nodes and internodes
   b. The *axillary buds* located above each leaf scar (an area where a leaf had previously been attached) are themselves embryonic shoots
      (1) The terminal buds inhibit the growth of the axillary buds, a phenomenon called *apical dominance*
      (2) Apical dominance concentrates the growth of a shoot at its tip
   c. The *terminal bud scale scars* are the remains of the terminal buds; the distance between a terminal bud and the terminal bud scar or between two terminal bud scars represents 1 year's growth

IX. Reproduction in Higher Animals

A. General information

1. Reproduction in higher animals (such as reptiles, birds, and mammals) involves the production of gametes (*gametogenesis*), which takes place in special reproductive organs called *gonads*
   a. In *spermatogenesis,* the male primary sex cells undergo reduction division to produce four haploid cells of approximately equal size; each male gamete (spermatozoon) has a long flagella and little cytoplasm
   b. In *oogenesis,* the female primary sex cells undergo reduction division to produce four haploid cells of unequal size
      (1) The first meiotic division produces one large cell and one small cell called the polar body
      (2) The second meiotic division results in two additional polar bodies from the original one and a third polar body from the larger cell
      (3) This uneven division conserves cytoplasm in one cell, allowing for larger storage of food once the cell is fertilized
2. Gonads produce sex hormones and gametes
   a. The male gonads, called testes or spermaries, produce sperm and male sex hormones
   b. The female gonads, called ovaries, produce the ova and female sex hormones
3. Fertilization is the mechanism by which higher animals reproduce
4. After fertilization occurs, the zygote goes through a period of embryonic development and tissue differentiation

B. Fertilization

1. *External fertilization* occurs when the eggs are laid by the female and fertilized by the male outside the female's body; it is characteristic of fish and amphibians and typically produces large numbers of zygotes
2. *Internal fertilization* occurs when the sperm are deposited in or near the female reproductive tract and the ovum and sperm unite within the female's body; it is characteristic of birds and mammals and typically produces fewer zygotes
3. During mating, many sperm surround the egg cell (ovum), but only one can penetrate and unite with the egg's nucleus; after fertilization, a membrane forms around the egg to keep other sperm out

C. Embryonic development

1. Through a series of mitotic divisions, the fertilized egg (zygote) develops into an embryo
   a. In ***viviparous*** animals, the embryos develop and are nourished internally, a pattern typical of mammals
   b. In ***oviparous*** animals, eggs are laid and develop outside the female's body; birds, amphibians, and fish reproduce this way
   c. In ***ovoviviparous*** animals, eggs remain in the female's oviducts until hatched, a pattern seen in some fish and snakes
2. The initial series of cell divisions in the zygote is called *cleavage;* each cleaved cell is called a *blastomere*

3. The *morula,* a solid ball of cells, is produced by continued cleavage of the blastomeres
4. The cells eventually arrange in a single layer around the *blastocoel,* a fluid-filled cavity that develops in the center of the morula; this stage is called the *blastula*
5. The blastula is transformed into a cup-shaped embryo with two cell layers, called the *gastrula*
   a. This transformation is accomplished through *gastrulation,* a type of invagination, or punching in, of the cells
   b. The outer layer of the gastrula is called the *ectoderm,* and the inner layer is called the *endoderm;* a third, or middle, layer called the *mesoderm* soon develops

D. Differentiation

1. After the gastrula is formed, the three cell layers begin to differentiate
2. The ectoderm becomes skin, nerve tissue, hair, nails, and eye lenses
3. The mesoderm becomes muscle, bone, gonads, and the excretory and circulatory systems
4. The endoderm becomes the digestive system, pancreas, liver, thyroid, lungs, and bladder

X. Systems for Sex Determination

A. General information

1. Most animals have two separate sexes —male and female
2. In many plants and animals, such as the earthworm and the garden snail, a single individual produces both male and female gametes; such plants and animals are described as *monoecious*
3. Each separate-sex species has a mechanism for determining an individual's sex at the time of fertilization
4. In most cases (bees and ants are the exception), sex is determined by the combination of sex chromosomes in the fertilized egg

B. X-Y system

1. In this system, used by humans and most mammals, females have two X chromosomes (XX) and males have one X and one Y chromosome (XY)
2. Because they have two different chromosomes, males are the *heterogametic* sex —that is, their chromosomes determine the sex of the offspring

C. X-O system

1. This system is used by such insects as grasshoppers, crickets, and roaches
2. Females have two X chromosomes (XX) and males have only a single X chromosome designated XO

D. Z-W system

1. Offspring of birds, fish, and some insects (including butterflies and moths) are determined by the Z-W system
2. Females have one Z and one W chromosome (ZW); males have two of the same sex chromosomes (ZZ)

3. In this system, females are the heterogametic sex

E. Haplodiploidy

1. Bees and ants use this system, in which there are no sex chromosomes
2. Sex is determined by whether the egg is fertilized or not
3. Females develop from fertilized eggs and are diploid; males develop from unfertilized eggs (parthenogenesis) and are haploid

XI. Chromosomal Alterations

A. General information

1. Errors in meiosis or exposure of a cell to radiation or chemicals (mutagens) can adversely affect the chromosomes
2. Because chromosomes contain genes, which determine inherited characteristics, chromosomal alterations can have a significant impact on the future of the organism and even the species
3. The most common forms of chromosomal alterations include aneuploidy, polyploidy, and chromosomal fragmentation

B. Aneuploidy

1. This alteration results from nondisjunction during meiosis (that is, when both members of a pair of homologous chromosomes or both sister chromatids fail to move apart properly); it results in one gamete receiving two of the same chromosome and the other gamete receiving none
2. Fertilization by one of these gametes produces a zygote with an abnormal number of chromosomes
3. Variations of aneuploidy include trisomy and monosomy
   a. *Trisomy* results from the presence of an extra chromosome in an otherwise diploid cell
      (1) Klinefelter's syndrome is a trisomy caused by the presence of an extra X chromosome (XXY); those with this syndrome are male, but they have abnormally small testes and are sterile
      (2) An extra X chromosome in females (XXX) results in limited fertility and possible mental retardation
      (3) Trisomy 21, or Down's syndrome, results in a complex of physical deformities and mental retardation
   b. *Monosomy* results from the absence of a chromosome (Turner's syndrome is a condition in which only one X chromosome is inherited; the only known human monosomy, Turner's syndrome results in a female characterized by immature sex organs at adolescence and no secondary sex characteristics)

C. Polyploidy

1. This alteration occurs when more than two complete sets of chromosomes are inherited (inheritance of three sets of chromosomes is called *triploidy;* four sets, *tetraploidy*)
2. Polyploidy is common in plants but rare in animals; its occurrence in plants can lead to the formation of new species

D. Chromosomal fragmentation

1. This alteration occurs when only a fragment of a chromosome is lost, duplicated, or changed in some way
2. Variations include deletion, duplication, and translocation
   a. *Deletion* occurs when a fragment of a chromosome without a centromere is lost during cell division
      (1) The chromosome from which the fragment originated will then have a deficient number of genes
      (2) In humans, a deletion of chromosome 5 results in a disorder called *cri du chat,* which is characterized by mental retardation, an unusually small head, and a cry that sounds like that of a cat
   b. *Duplication* occurs when a fragment of one chromosome attaches itself to a homologous chromosome; although no genetic material is lost initially, after many mitotic divisions the absence of this material (or addition of material on those chromosomes receiving duplicated fragments) can cause problems for subsequent generations
   c. *Translocation* occurs when a chromosomal fragment attaches to a nonhomologous chromosome; in humans, chronic myelogenous leukemia results when a portion of chromosome 22 switches with a small fragment from the tip of chromosome 9

Study Activities

1. Compare sexual and asexual reproduction, providing an example of each for both plants and animals.
2. Describe what changes the chromosomes undergo in each step of mitosis and meiosis.
3. List the main reproductive organs of a flowering plant; identify their counterparts in a higher animal.
4. Outline the similarities among the X-Y, X-O, and Z-W systems of sex determination.
5. Sketch what happens during meiosis to produce aneuploidy.

5

The Basis of Inheritance

Objectives

After studying this chapter, the reader should be able to:
- Discuss Mendel's contribution to the study of genetics.
- Identify four situations that complicate genotype expression.
- Describe the structure and function of DNA and RNA.
- Differentiate between transcription and translation of DNA.
- Define genetic mutation and list the three ways in which it can occur.

I. Mendelian Genetics

A. General information

1. Much of what scientists know about *genetics*—the study of inheritance—can be traced back to the work of Gregor Mendel who, in the 1860s, demonstrated that parents pass to their offspring discrete heritable factors (genes) that retain their individuality from generation to generation
2. Genetics focuses on three areas: transmission of genes, expression of genes, and the genes' capacity to change (mutate)
3. Mendel's studies focused on the inheritance patterns of common garden peas, which can be either cross-fertilized or self-fertilized
4. To conduct his experiments, Mendel used several strains of peas with contrasting characteristics and studied them from generation to generation
   a. A *contrasting characteristic* is an easily distinguishable trait, such as the color of a pea pod (pea pods are either green or yellow—the only two clear alternatives for this trait)
   b. The contrasting characteristics selected by Mendel included seed texture, seed color, flower color, flower position, stem length, pod shape, and pod color
   c. Through experimentation, Mendel predicted which characteristics would appear in the offspring of crossbred plants; although these experiments predated the discovery of chromosomes and genes, certain inheritance patterns were evident and can be described in modern scientific terms
      (1) Homologous pairs of chromosomes contain pairs of genes called ***alleles;*** for example, a pea plant that is pure (shows the same trait after many generations) for the "tall" trait could be represented as having two TT alleles in every cell; a pea plant pure for the "short" trait would have two tt alleles

(2) Two individuals with two identical genes of an allelic pair, such as TT or tt, are *homozygous* for a particular trait; two individuals having two unlike genes of an allelic pair, such as Tt, are *heterozygous* for the trait

(3) The description of two genes in an allelic pair is called the ***genotype,*** whereas the appearance caused by the presence of both genes is called the ***phenotype*** (TT, tt, and Tt are genotypes, whereas "tall" and "short" are phenotypes)

5. Based on the findings of his experiments, Mendel identified three laws of inheritance —the *law of dominance,* the *law of segregation,* and the *law of independent assortment*

B. Law of dominance

1. When organisms having pure contrasting traits are crossed, the offspring (called hybrids) will show only one of these traits
2. The *dominant* trait, which appears in the offspring, takes precedence over the *recessive* trait, which does not appear
3. If a male (TT) and female (tt) pea plant are cross-fertilized, several combinations of alleles occur, as shown in Punnett square 1 (a Punnett square is a diagram of the possible genotypes that can result from a genetic cross)
   a. Mendel observed that this cross-fertilization resulted in all tall offspring; thus T is the dominant allele, and t is the recessive allele
   b. Whenever a dominant allele and a recessive allele are present together in a cell, the trait carried by the dominant allele is expressed in the first (F1) generation

C. Law of segregation

1. When large numbers of hybrids are crossed, the heritable factors segregate and recombine to produce dominant and recessive offspring in a 3:1 ratio
2. If two pea plants from the preceding F1 generation are cross-fertilized (each pea from the preceding F1 generation has the genotype Tt), then the genotypes shown in Punnett square 2 would occur in the F2 generation
3. The second (F2) generation offspring would be 75% tall (TT or Tt) and 25% short (tt), a 3:1 ratio

D. Law of independent assortment

1. Traits are inherited independently and are not affected by each other
2. If two plants pure for two contrasting traits (such as tall plants with yellow seeds and short plants with green seeds, where green is recessive to yellow) are crossed, the combinations in Punnett square 3 (see page 52) would occur
   a. All offspring of the F1 generation, called *dihybrids,* are tall with yellow seeds (TtYy)
   b. If two dihybrid pea plants from the F1 generation are crossed, the F2 generation offspring would appear in a ratio of 9:3:3:1 —9 tall and yellow to 3 short and yellow to 3 tall and green to 1 short and green (see *Punnett Square 4,* page 52)

E. Determining homozygous and heterozygous genotypes

1. A pea plant that has green pods (green pod color in pea plants is dominant, yellow pod color is recessive) could be homozygous or heterozygous

2. The appearance of the offspring of a cross between a pea plant that has a green pod and one having a yellow pod will reveal the genotype of the green pod parent
   a. If all F1 progeny have green pods, the original pea plant with green pods is homozygous (GG)
   b. If both green and yellow progeny appear in the F1 generation, the original pea plant with green pods is heterozygous (Gg)

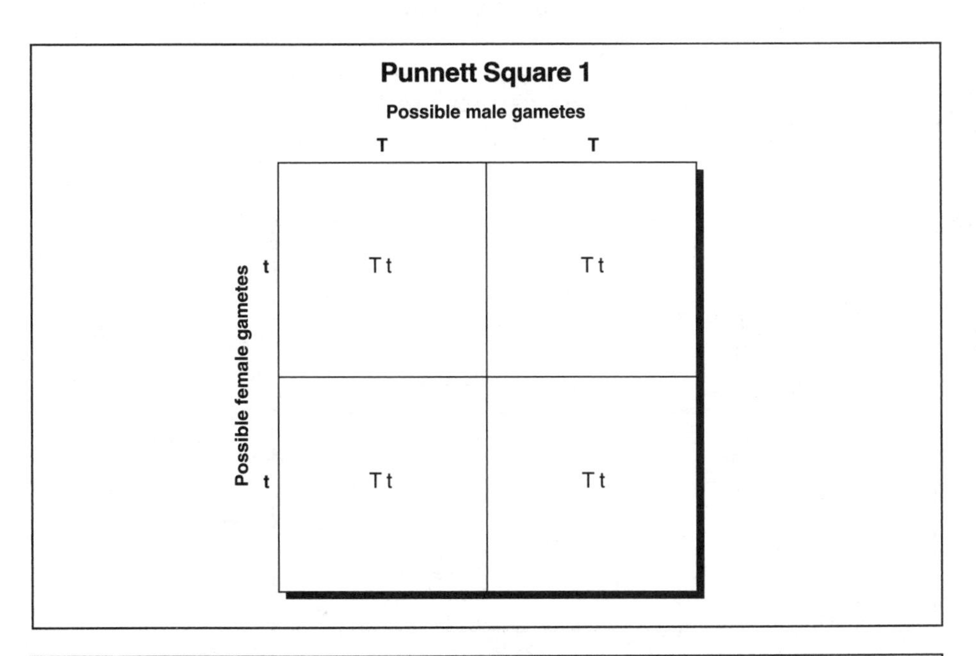

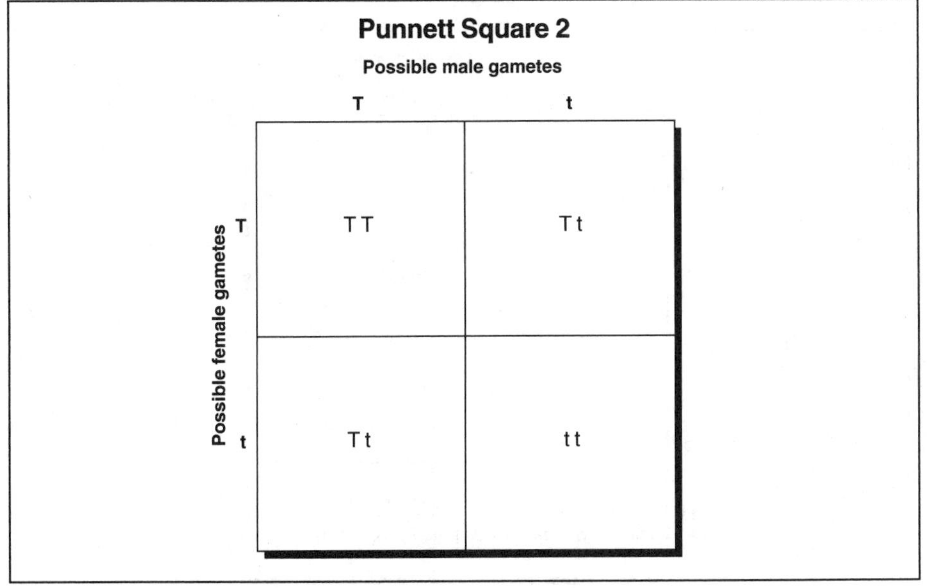

3. The breeding of a recessive homozygote with an organism of unknown genotype, called a testcross, was devised by Mendel

F. The influence of Mendel's work

1. Mendel's studies were conducted in the 1860s but did not influence the scientific community until after his death
2. In the early 20th century, biologists began to see parallels between the behavior of Mendel's heritable factors and the behavior of chromosomes

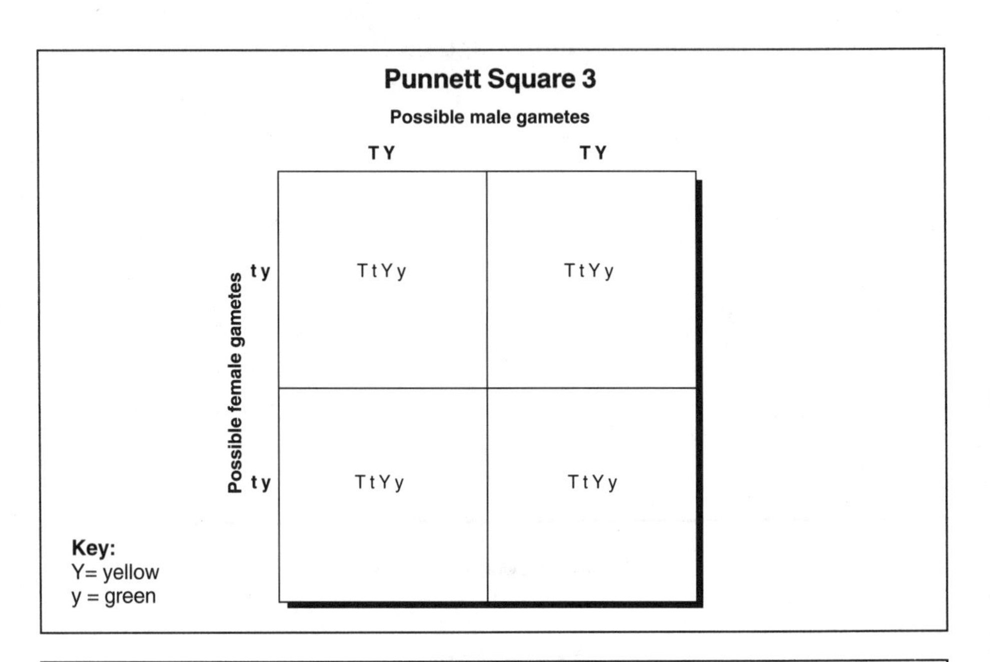

Punnett Square 4

	T Y	T y	t Y	t y
T Y	T T Y Y tall yellow	T T Y y tall yellow	T t Y Y tall yellow	T t Y y tall yellow
T y	T T y Y tall yellow	T T y y tall green	T t y Y tall yellow	T t y y tall green
t Y	t T Y Y tall yellow	t T Y y tall yellow	t t Y Y short yellow	t t Y y short yellow
t y	t T y Y tall yellow	t T y y tall green	t t y Y short yellow	t t y y short green

Key:
T= tall
t = short
Y= yellow
y= green

3. In 1902, Walter S. Sutton, Theodor Boveri, and others described a chromosomal theory of inheritance that asserted that Mendelian genes are located on chromosomes
4. Shortly thereafter, Thomas Hunt Morgan was the first to associate a specific gene with a specific chromosome
5. By the mid-20th century, scientists realized that genes were made up of deoxyribonucleic acid (DNA)
6. In 1953, James D. Watson and Francis H.C. Crick described the molecular structure of DNA and how this molecule could serve as the chemical basis of inheritance
7. The Watson-Crick model permitted a better understanding of how genes are expressed and how they can change over time

II. Complications from Genotype to Phenotype

A. General information

1. Mendel's laws hold true for all organisms only as generalizations and only in simple cases
2. They are not comprehensive enough to describe more complex inheritance patterns, such as intermediate inheritance, multiple alleles, pleiotropy, penetrance and expressivity, epistasis, polygenic inheritance, and linkage

B. Intermediate inheritance

1. Also called incomplete dominance, intermediate inheritance occurs when a characteristic is neither dominant nor recessive
2. Intermediate inheritance results in heterozygous offspring that are intermediate in appearance between the two homozygous expressions of the trait
   a. In snapdragons, neither red nor white flowers are dominant; red flowers occur in plants with the genotype RR, white flowers in plants with the genotype rr, and pink flowers in plants with the genotype Rr
   b. In humans, familial hypercholesterolemia (a disorder in which blood cholesterol levels are extremely high) is inherited through intermediate inheritance
      (1) People who are homozygous recessive (two recessive [abnormal] genes) have six times the normal blood cholesterol level
      (2) People who are heterozygous (one normal gene and one recessive gene) have twice the normal blood cholesterol level
      (3) People who are homozygous dominant (two normal genes) also have twice the normal blood cholesterol level

C. Multiple alleles

1. This situation occurs when some genes exist in more than two allelic forms
2. The ABO blood types in humans are an example of multiple alleles
   a. The three alleles in this blood group system are I^A, I^B, and i; thus six genotypes are possible
   b. Both I^A and I^B are dominant to i, making four different phenotypes possible

c. The alleles I^A and I^B control the presence of a different antigen on the surface of red blood cells; the allele i does not control the presence of any antigen
d. If the A antigen is present, then the blood serum contains circulating anti-B antibodies; if the B antigen is present, then the blood serum contains circulating anti-A antibodies; if both A and B antigens are present, then neither antibody is present; if no antigen is present on the surface of the red blood cell, then the serum contains anti-A and anti-B antibodies (see *Human Blood Types)*

D. Pleiotropy

1. The quality of a single gene to produce multiple phenotypic effects is known as *pleiotropy*
2. In Siamese cats, a single gene can cause both abnormal pigmentation and cross-eyes
3. In humans, sickle cell anemia is due to a single gene that causes a complex set of symptoms

E. Penetrance and expressivity

1. *Penetrance* is the proportion of individuals who exhibit an expected phenotype; penetrance is 100% if the genotype mandates the phenotype in all cases
2. Incomplete penetrance occurs when not all organisms with the same genotype express identical phenotypes; for example, of the humans who inherit the dominant allele for retinoblastoma (a type of eye tumor), not all of them develop the disease
3. The same gene also can be *expressed* in different individuals to varying degrees; in the case of retinoblastoma, the severity of the disorder varies among those who develop the tumor

F. Epistasis

1. Epistasis is a situation in which one gene interferes with the expression of another independently inherited gene
2. The first gene is said to be epistatic to the second gene
3. The inheritance of flower color in the sweet pea is an example of epistasis
    a. The dominant gene P causes purple flowers, whereas another independently inherited gene controls a metabolic pathway for color pigmentation
    b. If this latter gene, which is recessive, is also homozygous, the flowers will be white, regardless of the P gene

G. Polygenic inheritance

1. Many traits result from the additive effect of two or more genes (the opposite of pleiotropy)
2. In humans, skin pigmentation is controlled by at least three separately inherited genes
    a. The presence of all three genes in their dominant allelic form produces the darkest skin
    b. The presence of all three genes in their recessive allelic form produces the lightest skin
    c. Combinations of the three genes produce skin of varying shades

Human Blood Types

BLOOD GROUP PHENOTYPE	GENOTYPES	ANTIBODIES PRESENT IN BLOOD SERUM
A	$I^A I^A$ or $I^A i$	Anti-B
B	$I^B I^B$ or $I^B i$	Anti-A
AB	$I^A I^B$	None
O	ii	Anti-A and Anti-B

3. The height trait also is thought to be inherited in a polygenic fashion

H. Linkage

1. Because chromosomes are passed on from cell to cell as indivisible units, the genes on the same chromosome are often inherited (linked) together
2. In *Drosophila melanogaster,* the common fruit fly, the genes for black body and short wings are linked, thus these two traits are inherited together

III. DNA

A. General information

1. DNA molecules in the cell's chromosomes contain the code for various inherited characteristics
2. The information contained within this genetic code directs the formation of enzymes and proteins involved in the cell's activities
3. Three lines of evidence have proven that genes are made up of DNA
   a. In transformation experiments, scientists successfully transferred DNA from one bacterium to another, changing the heredity patterns of the bacteria
   b. In experiments involving bacteriophages —bacteria-attacking viruses (phages) that are capable of injecting their own DNA into bacteria —scientists were able to replicate viral DNA within the bacteria and produce more new bacteriophages than new bacteria
   c. In transduction experiments, scientists introduced a bacteria-attacking virus into bacterial cells, resulting in the production of new viral offspring containing some of the bacterial cells' DNA

B. DNA structure

1. Each DNA molecule is comprised of thousands of basic subunits called *nucleotides*
2. Each nucleotide consists of a 5-carbon sugar (deoxyribose), a phosphate, and one of four nitrogen bases (adenine, guanine, cytosine, or thymine)
3. The arrangement of these nucleotides was described by James D. Watson and Francis H.C. Crick based on analysis of X-ray diffraction data supplied by Maurice H.F. Wilkins (see *Watson-Crick Model of DNA,* page 57)

C. Characteristics of DNA

1. Two nucleotide strands are arranged in a ladder pattern, which is then twisted to form a double spiral, or helix
2. The upright parts of the ladder are comprised of deoxyribose and phosphate; the rungs are comprised of paired nitrogen bases (adenine with thymine and guanine with cytosine)
3. The nitrogen bases are bound by a relatively weak hydrogen bond

D. Replication of DNA

1. When chromosomes replicate during mitosis and meiosis, each DNA molecule also replicates by forming an exact copy of itself
2. DNA replication begins with the opening of the DNA helix at specific points, called the origins of replication
3. Several types of proteins are involved in opening the double helix
   a. *Helicases* are enzymes that unwind the helix
   b. *Single-strand binding proteins* keep the strands separated
   c. *Topoisomerases* keep the strands from tangling
4. After separation, each parent DNA strand can serve as a template for the synthesis of two new DNA strands
5. The numerous free nucleotides in the cell move into position to bond with a matching nucleotide of the parent strand (adenine with thymine and guanine with cytosine); the enzyme DNA polymerase acts as a catalyst in this process

IV. Ribonucleic Acid

A. General information

1. Ribonucleic acid (RNA), located in the cytoplasm of the cell, communicates instructions for making the enzymes and proteins necessary for DNA synthesis
2. RNA is categorized according to three types: messenger, transfer, and ribosomal
   a. *Messenger RNA* (mRNA) carries the genetic code from the DNA in the nucleus to the ribosomes (site of protein synthesis) in the cytoplasm
   b. *Transfer RNA* (tRNA) transfers amino acids in the cytoplasm to the ribosomes; more than 20 different types of transfer RNA exist—one for each type of amino acid
   c. *Ribosomal RNA* (rRNA), located in the ribosomes (which are made up of protein and rRNA), lines up the amino acids in the sequence dictated by the information contained within the mRNA

B. RNA structure

1. The structure of RNA is similar to that of DNA, with some exceptions
2. Its 5-carbon sugar is ribose, not deoxyribose
3. Its four nitrogen bases are adenine, cytosine, guanine, and uracil
4. It is a single-stranded, not a double-stranded, molecule

Watson-Crick Model of DNA

Deoxyribonucleic acid (DNA) contains the hereditary code of an organism. Each DNA molecule is comprised of thousands of subunits called nucleotides, each of which consists of three parts: a five-carbon sugar (deoxyribose), a phosphate group, and a nitrogen group. The nitrogen group of each nucleotide contains one of four different bases: adenine, guanine, cytosine, or thymine.

The nucleotides within each DNA molecule are arranged in a sequenced strand, with the phosphate of one nucleotide bound to the sugar of the next nucleotide; the nitrogen group is always situated to the side of the sugar in each nucleotide. Each molecule consists of two such strands held together by a weak hydrogen bond that connects the nitrogen groups of one strand with those of the other. This double strand is coiled into a helix configuration, which when uncoiled resembles a ladder.

According to the Watson-Crick Model of DNA:

- two strands of nucleotides are arranged in a ladder-type organization
- this ladder is then twisted into a double spiral or helix (double helix)
- the upright parts of the ladder are made up of deoxyribose and phosphate
- the rungs of the ladder are made up of paired nitrogen bases, which are always arranged according to the same configuration (that is, adenine with thymine and guanine with cytosine)
- the bond between the nitrogen bases is a relatively weak hydrogen bond.

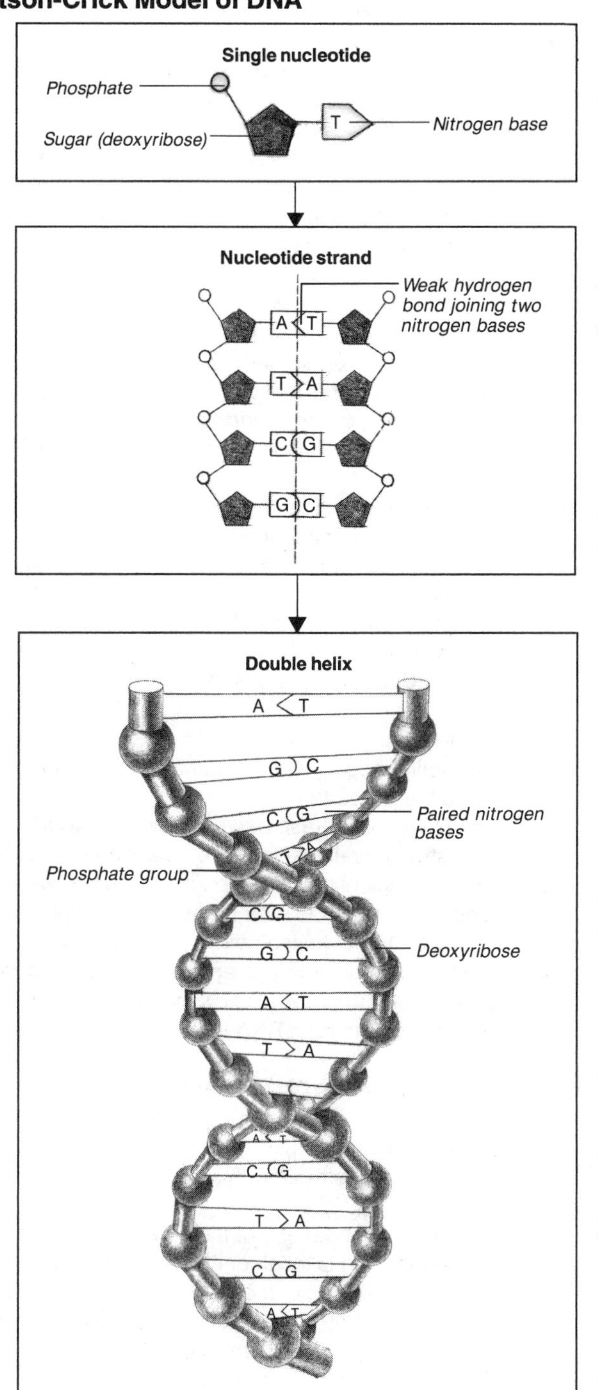

V. Protein Synthesis

A. General information

1. DNA (which forms the genes) controls the hereditary characteristics of an organism by specifying the sequence of polypeptides when proteins are formed
2. One gene codes for one polypeptide
3. The two processes involved in protein synthesis are transcription and translation
   a. *Transcription,* the transfer of information from a DNA molecule to an RNA molecule, occurs in the nucleus of eukaryotic cells
   b. *Translation,* the transfer of information from an RNA molecule to a polypeptide, occurs in the cytoplasm of eukaryotic cells

B. Genetic code

1. The transfer of information from gene to protein is based on the triplet code, a sequence of three nucleotides in DNA or RNA
2. Each sequence, called a ***codon,*** specifies a particular amino acid or termination signal (a termination signal is a special codon that signals the end of transcription)
3. In DNA, the triplet code consists of the four letters A, T, G, and C, which correspond to one of the four nucleotides (adenine, thymine, guanine, and cytosine); in RNA the four-letter code is A, U, G, and C (in RNA, uracil is substituted for thymine)
4. Each amino acid is specified by a codon; for example, the amino acid phenylalanine is specified by the RNA codon UUU, the amino acid alanine by the RNA codon CCG or UCG, and the amino acid histidine by the RNA codon ACG

C. Transcription

1. Transcription begins when the enzyme RNA polymerase binds to a piece of DNA called the *promoter*
2. As the RNA polymerase moves along, it unwinds the DNA helix, making one DNA strand available to serve as a template for the synthesis of mRNA
3. Free RNA nucleotides in the nucleus line up next to the appropriate DNA nucleotides in the exposed strand (the free nucleotides follow the rule of nitrogen bases, that is, cytosine pairs with guanine and adenine pairs with uracil)
4. When the RNA polymerase encounters a special sequence of DNA nitrogen bases called the *terminator,* it stops transcribing and both the RNA polymerase and the completed mRNA molecule are released

D. Structure and function of transfer RNA

1. tRNA molecules link the appropriate amino acid with the appropriate mRNA codon
2. They carry out their role with a special multidimensional structure that resembles a cloverleaf
   a. At the cloverleaf end of the tRNA, a sequence of three nucleotides called the ***anticodon*** is located; this sequence matches a codon triplet on the mRNA
   b. At the other end of the tRNA, a receptor site at which an amino acid can attach is located; each tRNA attaches to only one amino acid, and the anticodon determines which one

3. Amino acid-activating enzymes facilitate the attachment of the amino acid to the tRNA

E. Translation

1. Translation requires four separate steps: initiation, elongation, translocation, and termination
2. During *initiation,* mRNA begins the translation process by binding to a small ribosomal subunit
    a. A special initiator tRNA recognizes the mRNA's codon and binds to it
    b. A large ribosomal subunit (made up of proteins and rRNA) then joins the entire complex
        (1) The site on the ribosome at which the initiator tRNA is bound to the mRNA is called the P site
        (2) The site immediately next to the P site is called the A site
3. During *elongation,* which occurs after initiation is complete, amino acids are added one by one to the initial amino acid on the initiator tRNA
    a. The mRNA in the A site binds with a second complementary tRNA molecule
    b. The enzyme peptidyl transferase (which is part of the large ribosomal subunit) catalyzes the formation of a peptide bond between the amino acid in the P site and the amino acid carried by the second tRNA molecule
4. During *translocation,* the tRNA on the P site dissociates from the ribosome and the tRNA in the A site is translocated to the P site along with the mRNA to which it is attached
    a. The movement of the tRNA brings the next mRNA codon into the A site
    b. Another incoming tRNA attaches to this codon, and the process continues
5. During *termination,* a termination codon reaches the A site
    a. A protein called the *release factor* then binds to the A site
    b. The release factor causes the completed polypeptide to dissociate from the tRNA and ribosome

VI. Genetic Mutations

A. General information

1. A ***mutation***—a variation in the nucleotide sequence of DNA—provides genes with the capacity for change, an important link between genetics and evolution
2. A mutation can affect a large region of a chromosome or a single nucleotide pair (the latter is called a *point mutation*)
3. Point mutations can occur as a result of a base pair substitution or a base pair insertion or deletion
4. *Mutagenesis,* the process by which mutations are created, may occur as a result of spontaneous mutation or the presence of a mutagen or transposon
    a. *Spontaneous mutations* occur when errors arise during DNA replication, repair, or recombination
    b. Physical agents (such as ultraviolet radiation) or chemical agents also can alter the genetic code; such agents are called *mutagens* (for example, ultraviolet radiation can break DNA's double strands, resulting in chromosomal rearrangements and deletions; chemical agents can cause

additions to or deletions from chemical groups, leading to mispairing during DNA synthesis)

c. *Transposons*—transposable elements or mobile segments of DNA that move from one gene to another—can act as mutagens by disrupting the function of the gene into which they move

B. Base pair substitutions

1. This category of point mutation occurs when one nucleotide and its partner are replaced
2. Sickle cell anemia is an example of a base pair substitution that has a disastrous effect in humans
   a. In this disorder, the gene that codes for hemoglobin has a single base substitution in one of its codons; the codon for glutamic acid (CTT) is changed to the codon for valine (CAT)
   b. The resulting hemoglobin molecule has valine substituted for glutamic acid, which produces the abnormal hemoglobin that causes the disease

C. Base pair insertions or deletions

1. This category of point mutation occurs when one or more nucleotide pairs in a gene are either inserted or deleted
2. Base pair insertions or deletions tend to do more harm than base pair substitutions because all the nucleotides downstream from the insertion or deletion become grouped into inappropriate codons

VII. Genetic Disorders in Humans

A. General information

1. Mutated genetic material can cause inherited disorders
2. A *dominantly inherited* disorder is caused by a single mutated allele; an individual who receives one mutated allele from either parent exhibits the disorder
3. A *recessively inherited* disorder is caused by two copies of the mutated allele; an individual must receive two mutated alleles, one from each parent, to exhibit the disorder

B. Dominantly inherited disorders

1. *Achondroplasia* involves inadequate bone formation resulting in a specific kind of dwarfism
2. *Huntington's chorea* is characterized by the progressive deterioration of the nervous system beginning in mid-life
3. *Syndactyly* results in webbed fingers and toes
4. *Osteogenesis imperfecta* causes brittle bones
5. *Alzheimer's disease* is a type of dementia most often seen in older people and characterized by the degeneration of brain tissue and the loss of intellectual functioning
6. *Galactosemia* is a disorder of carbohydrate metabolism that can cause malnutrition in infants
7. *Congenital cataracts* cause opacity of the lens of the eye

C. Recessively inherited disorders

1. *Albinism* is characterized by the absence of pigment in the eyes and the skin
2. *Cystic fibrosis* results in the presence of excessive mucus secretions from the pancreas, lungs, and other organs, leading to blockage of the digestive tract, cirrhosis of the liver, and pneumonia and other infections (cystic fibrosis is the most common lethal genetic disease in the United States, striking 1 in 2,500 whites)
3. *Tay-Sachs disease* causes a breakdown in normal lipid metabolism, leading to an accumulation of lipids in the brain; death results from the abnormal functioning of the brain cells (Tay-Sachs disease is most common in Ashkenazi Jews; the incidence in this population is 1 in 3,600 births)
4. *Phenylketonuria* is characterized by the inability of the liver to produce the enzyme necessary to convert phenylalanine to tyrosine, resulting in an accumulation of phenylalanine and its metabolites in the brain; phenylketonuria produces severe mental retardation in infants
5. *Thalassemia* and *sickle cell anemia* are types of severe anemia

VIII. Recombinant DNA

A. General information

1. Recombinant DNA, a form of DNA derived from two different sources and recombined in vitro, has led to many new discoveries in molecular biology and the formation of a new biotechnology industry called genetic engineering
2. Recombination involves extracting a section of DNA from one area and relocating it to another area
3. DNA from different species ordinarily never recombines; however, recombinant DNA technology makes this type of gene manipulation possible
4. The chief tool of the recombinant DNA technologist is a special class of enzymes called *restriction enzymes*
   a. Found in bacteria, restriction enzymes can split a DNA strand into fragments that have "sticky ends"
   b. The sticky ends consist of a single strand of DNA that readily binds with any available complementary base; when these fragments touch a DNA strand from another organism, they stick to it, producing recombinant DNA

B. Recombination

1. Recombinant DNA produced in vitro can be introduced into bacteria, where it reproduces and expresses the desired trait
2. To accomplish this, a piece of DNA must be fragmented using restriction enzymes
   a. The desired fragment or combination of fragments (now called the foreign gene) is isolated by electrophoresis
   b. The foreign gene is combined with a bacterial plasmid (a ring of bacterial DNA), which also has been spliced using restriction enzymes
   c. The plasmid, along with the foreign gene, is introduced into a bacterial culture, where it is taken up into live bacterial cells (the plasmid acts as a vector—that is, a means of moving the recombinant DNA from the test tube into the cell)

d. The bacteria, which reproduce rapidly, express the trait originally coded for by the foreign gene

C. Applications of recombinant DNA technology

1. Recombinant DNA technology has many current applications as well as applications for the future
2. Presently, many pharmaceutical products (including human insulin, erythropoietin, and human growth hormone) are being synthesized through recombinant DNA technology; the substances produced by these methods are identical to those produced in humans
3. Recombinant DNA techniques also can be used to produce large quantities of a specific protein molecule from the protein coat of a particular disease-carrying organism
   a. The protein coat (but not the entire microbe) can then be introduced back into the cells
   b. Serving as an antigenic stimulant, the protein coat is powerful enough to promote the production of large numbers of antibodies against the disease
4. In the future, replacing a defective gene with a copy of a normal gene may be possible with recombinant DNA technology
5. Recombinant DNA technology also may prove useful in agriculture by increasing the amount of nitrogen fixation carried out by bacteria

---

Study Activities

1. Define the three laws of Mendelian genetics, using a Punnett square to illustrate each law.
2. Determine the possible genotypes and phenotypes of the offspring from parents with blood types A and AB.
3. Identify the following nucleotide sequence as either DNA or RNA and provide the corresponding code: AAGCUGA
4. Compare dominantly and recessively inherited genetic disorders.
5. Outline the basic process of recombinant DNA technology.

6

Regulation and Control Mechanisms in Plants and Animals

Objectives

After studying this chapter, the reader should be able to:

- Discuss the concept of homeostasis and how it relates to the challenges of regulation and control in animals.
- Identify the five types of plant hormones and describe their effects.
- Describe two mechanisms of plant movement.
- Compare the invertebrate and vertebrate excretory systems.
- Define thermoregulation and the types used by endotherms.
- List seven glands of the vertebrate endocrine system, and describe the hormones that each produces.

I. Homeostasis

A. General information

1. Both plants and animals strive to maintain an internal steady state (equilibrium) in the face of a changing external environment
2. *Homeostasis* describes the physiologic adjustments made by organisms to maintain this steady state
3. During an animal's development, major changes in its internal environment are programmed to occur; for example, the balance of hormones in human blood is altered radically during puberty
4. Animals encounter three major challenges to the control of their internal environment
    a. They must control solute balance and the gain and loss of water *(osmoregulation)*
    b. They need to rid the body of nitrogen-containing wastes *(excretion)*
    c. They must maintain internal temperature *(thermoregulation)*

B. Control mechanisms

1. Most physiologic mechanisms for maintaining homeostasis are based on *negative feedback,* which allows an organism to reverse an internal change
    a. Negative feedback is analogous to a thermostat, which controls the temperature of a room

b. When the temperature falls below a set point, a sensor in the thermostat activates a switch that turns on the heater
c. When the temperature rises above a set point, a sensor in the thermostat activates a switch that turns off the heater
d. The hypothalamus in higher animals acts in this way to maintain body temperature at the optimum of 37° C; when the body temperature rises above 37°, the hypothalamus sends nervous impulses to sweat glands, which lower the body temperature through evaporative cooling; when the body temperature drops, the hypothalamus stops sending the nervous signals and the evaporative process stops

2. *Positive feedback* mechanisms enable an organism to amplify, rather than reverse, a change
   a. In human females during childbirth, the presence of the baby's head against the opening of the uterus stimulates uterine contractions that increase pressure on the uterine opening
   b. This increased pressure heightens the contractions, which further increases the pressure on the uterus, and so on
3. Both plants and animals have mechanisms for coordinating their overall internal responses to environmental stimuli
   a. Plants use a system of chemical substances (hormones) that can signal other cells within the organism and coordinate the organism's overall responses by controlling its growth and development
   b. Animals depend on hormones, pheromones, and local regulators to coordinate responses; animal hormones usually are part of a more elaborate endocrine system
   c. In higher animals, coordination of responses also may be controlled by a nervous system, which can react to environmental stimuli more quickly than an endocrine system can

II. Plant Hormones

A. General information

1. Plant growth and development are controlled by chemical substances (hormones) that affect the division, elongation, and differentiation of cells
2. These hormones are produced in minute quantities yet can have a significant effect on the plant
3. Plant hormones are classified according to five types: auxins, cytokinins, gibberellins, abscisic acid, and ethylene

B. Auxins

1. These hormones promote the elongation of young developing stems, root growth, differentiation and branching, apical dominance (concentration of growth at the tip of a plant shoot that results in partial inhibition of axillary growth), and fruit development
2. Auxins also play a role in phototropism (the turning of a plant in response to a light source) and gravitropism (the turning of a plant in response to gravity)
3. Found in the endosperm and produced by the seed embryo, auxins are synthesized primarily in the apical meristem of the shoot

4. From the apical meristem, auxins are transported downward from one cell to the next by means of ***polar transport***—that is, in a unidirectional mode, from the shoot's tip to its base
5. All auxins contain indoleacetic acid

C. Cytokinins

1. These hormones promote root growth and differentiation, stimulate cell division and growth, cause germination and flowering, and delay senescence (aging)
2. These substances are modified forms of adenine, a component of nucleic acids; they are produced in actively growing tissues —especially roots, embryos, and fruits —and are transported upward in the xylem sap
3. Cytokinins can slow the aging process of some plant organs by inhibiting protein breakdown and stimulating RNA and protein synthesis
4. They interact with auxins in controlling apical dominance
   a. Auxins and cytokinins are antagonistic; auxins prevent axillary bud growth, whereas cytokinins promote it
   b. The concentration of these two hormones at any one time determines whether axillary buds grow

D. Gibberellins

1. These hormones promote seed and bud germination, stem elongation, and leaf growth (they have little effect on root growth); stimulate flowering and fruit development; and affect root differentiation
2. Over 70 different gibberellins occur naturally in plants
3. They are produced in the meristems of apical buds, roots, and young leaves as well as by the seed's embryo
4. Gibberellins can cause rapid elongation of stems, a phenomenon called *bolting*
5. Both auxins and gibberrellins must be present for fruit development to occur

E. Abscisic acid

1. This hormone inhibits growth, closes the stomata during water stress, and induces dormancy
2. Abscisic acid is produced by leaves, stems, and green (unripe) fruit
3. It helps prepare the plant for winter by inhibiting cell division
4. In the spring, a seed germinates when the ratio of abscisic acid to gibberellins is altered so that the concentration of gibberellins is increased

F. Ethylene

1. Ethylene is unique among plant hormones because it is a gas that diffuses through the plant into the air spaces between the cells
2. Produced by the tissues of ripening fruit, stem nodes, and senescent leaves, ethylene promotes fruit ripening and leaf abscission (the dropping of leaves from deciduous trees)
3. Fruit ripening involves the degradation of cell walls (which softens the fruit) and a decrease in chlorophyll content (which causes the loss of green color)
4. Leaf abscission occurs in response to environmental stimuli, such as shorter days and cooler temperatures
   a. Leaf abscission involves a decrease in the production of auxins, which triggers an increase in the production of ethylene

b. Ethylene causes an increase in the enzymes that degrade the cell walls in the abscission layer of the leaf

III. Plant Movement

A. General information

1. Typically, plants are rooted to one location for life
2. However, plant movement can be accomplished through tropisms and changes in turgor
   a. *Tropisms* are growth and development responses that allow the plant to slowly adjust its position, either toward or away from a stimulus
   b. *Turgor* is the normal distention of the protoplasmic layer and wall of a plant cell by its fluid contents; changes in the turgor of specialized cells enable the plant to undergo reversible and rapid movement

B. Tropisms

1. When cells on opposite sides of the plant elongate at different rates, the plant moves
2. These tropisms can be positive (movement toward a stimulus) or negative (movement away from a stimulus)
3. The three types of tropisms are phototropism (movement in response to light), gravitropism (movement in response to gravity), and thigmotropism (movement in response to touch)
   a. In *phototropism,* light causes a redistribution of auxins in the growing shoot
      (1) The auxins become concentrated on the shaded side, increasing the rate of cell elongation on this side
      (2) The increased cell elongation on the shaded side bends the shoot toward the light
   b. In *gravitropism,* roots display positive gravitropism (downward growth), whereas shoots display negative gravitropism (upward growth)
      (1) Plants can distinguish up from down by the settling of statoliths (special plastids containing dense starch grains) in response to gravity
      (2) The statoliths trigger the accumulation of auxins and gibberellins in certain parts of the roots and stems
      (3) The hormones promote differential growth, which results in movement either away from or toward the pull of gravity
   c. *Thigmotropism,* seen in most vines and other climbing plants with tendrils, is the differential growth of a plant that enables it to coil around an object it touches

C. Changes in turgor

1. Activation of specialized, turgor-sensitive cells enables a plant to move rapidly
2. Such movement is typical of the compound leaf of the mimosa plant
   a. The plant's specialized pulvini cells (located at the joints of the leaf) lose potassium in response to touch; water then leaves the cells by osmosis, which causes the cells to become flaccid, the leaflets to fold together, and the entire leaf to drop
   b. It takes about 10 minutes for the cells to regain their original turgor

3. Many plants lower their leaves in the evening and raise them fully in the morning; called *sleep movement,* this change in position is accomplished by daily changes in the turgor of specialized cells
4. The Venus flytrap, the sundew, and the bladderwort have similar mechanisms for rapid movement by changes in turgor

IV. Circadian Rhythms and Photoperiodism

A. General information

1. Both plants and animals follow circadian rhythms, physiologic cycles that occur over a 24-hour period
2. In plants, circadian rhythms affect sleep movement and the opening and closing of stomata
3. In animals, they affect sleep-wake cycles and hormone levels
4. Only plants, however, have a direct physiologic response to day length, called *photoperiodism*

B. Circadian rhythm

1. The timing of the circadian rhythm depends on an organism's internal biological clock
2. Almost all eukaryotic organisms have a biological clock
   a. The clock is set by daily exposure to external cues from the environment, the most common is the light-dark cycle caused by the earth's rotation
   b. In the absence of daily external cues, the biological clock may deviate from a 24-hour period; this free-running period may last from 21 to 27 hours
   c. If the timing of the external cues changes, it takes a few days for the biological clock to reset itself
3. Scientists hypothesize that biological clocks operate at the cellular level, either in membranes or in the mechanisms of protein synthesis

C. Photoperiodism

1. Photoperiodism, a physiologic response to day length, occurs only in plants
2. Studies in the 1920s seemed to indicate that day length was the controlling factor in flowering; based on these studies, scientists described plants according to one of three types: short-day plants, long-day plants, and day-neutral plants
   a. *Short-day plants* require a shorter-than-normal period of light to flower; chrysanthemums and poinsettias, for example, typically flower in late summer or early fall, when daylight is shortest
   b. *Long-day plants* require a longer-than-normal period of light to flower; spinach, lettuce, and irises, for example, typically flower in late spring or early summer, when daylight is longest
   c. *Day-neutral plants* are unaffected by the period of light; tomatoes, garden peas, and dandelions fall into this category
3. In the 1940s, plant physiologists demonstrated that night length, not day length, was the critical factor in flowering; based on this knowledge, a short-day plant is actually a long-night plant and a long-day plant is actually a short-night plant

4. If a short-day plant is interrupted by a single burst of light during the night, it will not flower; similarly, a long-day plant can be forced to flower earlier by artificially shortening its night
5. The photoperiod is detected by the leaves, which apparently convey a message to the flowers; the hormone *florigen* may be involved in relaying this message
6. The plant measures the period of darkness through a special pigment called *phytochrome,* which is highly sensitive to red light or radiation (red radiation causes a structural change in phytochrome, which initiates a sequence of events that leads to flowering)

V. Osmoregulation

A. General information

1. Whether an animal inhabits land or fresh or salt water, its cells cannot survive a net gain or loss of water
2. *Osmoregulation*—a process that occurs only in animals—balances the intake and loss of water, which continuously enters and exits the cells across the plasma membrane
3. The direction of water movement is determined by the concentration of solutes (substances) in the solutions on either side of the plasma membrane

B. Water movement

1. Water always moves by *osmosis*—movement of a solvent through a semipermeable membrane into a solution of higher solute concentration that tends to equalize the solute concentration on both sides of the membrane
2. The solution with the greater concentration of solutes is considered *hyperosmotic,* whereas the one with the lesser concentration of solutes is *hypoosmotic;* solutions of equal solute strength are considered *isosmotic*
3. *Osmotic pressure* is the cell's ability to take up water from a pure reservoir (water with no solute dissolved in it); a cell with a high *osmotic concentration* (high level of solute in solution) has a high osmotic pressure
4. *Osmolarity,* the term used to express solute concentration, is key to the movement of water across cells
   a. If an animal cell without a rigid cell wall (such as a red blood cell) is placed in an isosmotic solution, no net movement of water across the plasma membrane occurs and the volume of the cell remains constant
   b. If the same cell is placed in a hyperosmotic solution (one that has an osmotic pressure greater than that of the cell), the cell loses water to the environment and shrivels, becomes crenated (shrinks) and, in some cases, dies
   c. If the cell is placed in a hypoosmotic solution (one that has an osmotic pressure less than that of the cell), water enters the cell faster than it can leave, and the cell swells and bursts (lyses)
   d. Cells without rigid cell walls that live in hyperosmotic or hypoosmotic environments must be specially adapted for the control of water balance

C. Maintaining water balance

1. *Osmoconformers* are animals that do not actively adjust their internal osmolarity

a. The osmolarity of their cells varies with the osmolarity of the external environment, thus the cell is always isosmotic with its surroundings
b. Most marine ***invertebrates*** are osmoconformers

2. *Osmoregulators* are animals whose cells are not isosmotic with the surrounding environment
a. Their cells must discharge water in a hypoosmotic environment or continuously take in water in a hyperosmotic environment
b. All freshwater animals, many marine animals, and most terrestrial animals, including humans, are osmoregulators

3. Most animal cells (except those of the simplest animals, such as sponges, cnidarians, and flatworms) are not in direct contact with the external environment but are bathed by an internal body fluid
a. In animals with an open circulatory system (one without a system of closed circulatory blood vessels, such as that found in insects), this internal body fluid is called *hemolymph*
b. In animals with a closed circulatory system (one with a well-defined network of blood vessels), the internal fluid is comprised of three fluid types —intracellular fluid (the cytosol of cells) and two extracellular fluids (blood plasma and interstitial fluid)

4. Most osmoregulators maintain water balance through the use of specialized epithelia called *transport epithelia*
a. Transport epithelia can regulate the transport of salt between the animal's internal fluid and its external environment
b. By regulating salt, they also regulate water movement, since water follows solute through osmosis
c. The molecular composition of the epithelium's plasma membrane determines the transport epithelia's specific osmoregulatory functions; for example, in the gills of saltwater fish, the transport epithelia pump salt out of the cells, whereas in the gills of freshwater fish, the transport epithelia pump salt into the cells
d. Besides maintaining salt and water balance in osmoregulators, transport epithelia also excrete nitrogenous wastes in many animals

VI. Excretion

A. General information

1. Metabolism produces toxic by-products that must be excreted by the body
2. The most toxic by-products are the nitrogenous wastes, such as ammonia, that result from the metabolism of proteins and nucleic acids
3. Most aquatic animals excrete ammonia unchanged; mammals and most adult amphibians convert it to urea first, whereas insects, birds, and some reptiles convert it to uric acid
4. Invertebrates control osmoregulation and excretion through a special system of tubules (which contain transport epithelia) scattered throughout their body
5. Vertebrates control osmoregulation and excretion through specialized tubules (which also contain transport epithelia) that are collected into compact organs called kidneys

B. Invertebrate excretory system

1. The invertebrate excretory system is characterized by three types of tubules: protonephridia, metanephridia, and malpighian
2. *Protonephridia* are a network of closed tubules that lack internal openings but contain an external opening called a nephridiopore
    a. Protonephridia are found in invertebrates without circulatory systems, such as flatworms
    b. In the flame cell system of the flatworm, interstitial fluid is pulled into a bulbous cell (the flame cell), which contains cilia that move the interstitial fluid along the length of the tubule
    c. The waste is then excreted through the nephridiopore
3. *Metanephridia* are a system of tubules that are open at both ends and closely associated with a network of capillaries
    a. This system is found in invertebrates with closed circulatory systems, such as earthworms (each worm segment contains metanephridia)
    b. In the earthworm, an opening into the coelomic (body) cavity, called a nephrostome, funnels fluid into and through the tubule
    c. Transport epithelia pump essential salts out of the fluid; these salts are reabsorbed by the surrounding capillaries
    d. The waste products are excreted from the worm's body through the nephridiopore
4. *Malpighian* tubules are found in insects, which have an open circulatory system
    a. In this system, the closed end of the tubule rests in the body cavity, where it is bathed by hemolymph; the other end of the tubule opens into the digestive tract
    b. Salt and nitrogenous wastes pass from the hemolymph into the tubule and are excreted into the digestive tract
    c. Before the wastes are excreted, much of the salt is pumped back into the body along with water (by osmosis)

C. Vertebrate excretory system

1. The *nephron,* a structure comprised of a renal tubule and its surrounding blood vessels, is the functional unit of the vertebrate excretory system
2. The nephron consists of the Bowman's capsule (the cup-shaped closed end of the renal tubule that receives filtrate from the blood); the glomerulus (a ball of capillaries surrounded by the Bowman's capsule); the proximal convoluted tubule; the loop of Henle; the distal convoluted tubule; and the collecting tubule
3. Nephrons regulate the blood's composition through filtration, secretion, and reabsorption
    a. During *filtration,* blood pressure forces fluid from the capillaries of the glomerulus into the renal tubule
        (1) Filtration is nonselective with regard to small molecules—that is, any substance small enough to be forced across the capillary wall by blood pressure enters the tubule
        (2) The filtrate that enters the renal tubule is a mixture of salts, glucose, vitamins, nitrogenous wastes, and many other small molecules
    b. During *secretion,* substances from the interstitial fluid surrounding the renal tubule pass into the tubule filtrate

(1) Secretion is selective and involves both active and passive transport mechanisms; for example, hydrogen ions may pass from the interstitial fluid into the tubule filtrate to maintain a constant pH level for body fluids

(2) Most secretion occurs in the proximal and distal convoluted tubules

c. During *reabsorption,* substances are selectively transported from the renal filtrate back into the plasma of the capillaries or the interstitial fluid

(1) Most reabsorption takes place in the convoluted tubules, the loop of Henle, and the collecting tubule

(2) Nearly all sugar, vitamins, and other organic nutrients and most of the water in the initial filtrate are reabsorbed

4. Through secretion and reabsorption, the kidney can control the concentration of various salts in the body fluids

VII. Thermoregulation

A. General information

1. An animal's metabolism is extremely sensitive to changes in internal temperature
2. Temperature affects the rate of cellular respiration, the effectiveness of enzymes, and the properties of cell membranes
3. To maintain an optimal internal temperature, animals exchange heat with the environment by conduction, convection, radiation, and evaporation
   a. *Conduction* is the direct transfer of heat between the body surface and the environment; heat is always conducted from a body of higher temperature to one of lower temperature; this type of heat transfer occurs when a body is cooled by being immersed in cool water
   b. *Convection* is the mass flow of air or liquid past a body surface; it occurs when a body is cooled by a breeze
   c. *Radiation* is the emission of electromagnetic waves produced by all objects warmer than absolute zero; radiation can transfer heat between objects not in direct contact with each other; it occurs when a body is warmed by the sun
   d. *Evaporation* is the loss of heat from the surface of a liquid; this type of heat transfer occurs when a body sweats

B. Thermoregulation in ectotherms

1. *Ectotherms,* animals that warm their bodies mainly from the environment, include invertebrates, fishes, reptiles, and amphibians
2. Ectotherms regulate body temperature through behavior; for example, a lizard will alternately sit in the sun and the shade to adjust its body temperature

C. Thermoregulation in endotherms

1. *Endotherms,* animals that derive most of their body heat from their own metabolism, include mammals and birds
2. Mechanisms for thermoregulation in endotherms include the rate of heat production, the rate of heat exchange, evaporative loss of heat, and behavioral responses

a. An animal can change its *rate of heat production* to adjust its internal temperature
   (1) When exposed to cold, an animal can double or triple its metabolic heat production by increased activity of skeletal muscles (either through movement or shivering)
   (2) The action of hormones also can trigger heat production (called nonshivering thermogenesis)

b. The *rate of heat exchange* between an animal and the environment can be adjusted in several ways, including vasodilation, vasoconstriction, hair or fur distribution, and body fat distribution
   (1) Vasodilation and vasoconstriction can alter the amount of blood flowing to the skin and thereby affect the amount of heat transferred by conduction, evaporation, and radiation
   (2) Hair and fur in mammals, feathers in birds, and fat in both birds and mammals can insulate against loss of body heat

c. *Evaporative loss of heat* can cause considerable cooling
   (1) Panting increases the evaporative loss from the respiratory tract
   (2) Sweat glands can increase the evaporative loss across the skin

d. *Behavioral responses,* such as relocation, basking in the sun, and finding cool damp areas in which to burrow, are other methods by which endotherms regulate temperature

D. Torpor

1. Torpor, which occurs in both ectotherms and endotherms, is an alternative physiologic state in which metabolism decreases and the heart and respiratory systems slow down
2. Torpor can take the form of hibernation, estivation, or diurnation
   a. *Hibernation* is a state in which body temperature is maintained at a lower-than-normal level, which enables the animal to withstand long periods of cold temperatures and decreased food supplies; many small mammals, such as insectivores and rodents, hibernate during the winter
   b. *Estivation* is a state in which metabolism slows and activity decreases, which enables the animal to survive long periods of elevated temperatures and decreased water availability
   c. *Diurnation* is a state similar to hibernation and estivation but lasts for a shorter period; many bats experience diurnation during daylight hours

VIII. Chemical Coordination of Responses in Animals

A. General information

1. All animals coordinate their metabolic responses through chemical signals
2. Hormones, pheromones, and local regulators serve as the chemical messengers

B. Hormones

1. These molecules are synthesized and secreted by specialized cells called *endocrine glands*

2. Released by the endocrine glands directly into the circulatory system, ***hormones*** travel to an area of the body called the target organ, where they activate target cells to elicit a specific biological response
3. Chemically, hormones are either steroids (such as sex hormones) or peptides (such as epinephrine)
4. The human body produces over 50 different hormones

C. Pheromones

1. These small, volatile molecules, which are easily dispersed into the environment, act as communication signals between animals of the same species
2. Minute quantities of a ***pheromone*** can have a potent effect
3. A pheromone produced by the female gypsy moth can elicit appropriate sexual responses in male gypsy moths

D. Local regulators

1. These chemical messengers affect target cells adjacent to or near their point of secretion
2. ***Neurotransmitters***, growth factors, and prostaglandins act as local regulators
   a. *Neurotransmitters* carry information from one neuron to another or from a neuron to a target cell, such as a muscle
   b. *Growth factors* regulate development within the animal's body; certain growth factors must be present in the extracellular environment for specific types of cells to grow and develop (for example, a protein called nerve growth factor is necessary for the development of axons)
   c. *Prostaglandins* are modified fatty acids derived from lipids in the plasma membrane and released into the interstitial fluid
      (1) Specific prostaglandins can have opposite local effects (for example, prostaglandin E causes blood vessels in the lungs to dilate while prostaglandin F causes the same blood vessels to constrict)
      (2) Other prostaglandins cause uterine contractions, induce fever and inflammation, and intensify the sensation of pain (aspirin may exert its effects by inhibiting prostaglandin secretion)

IX. Mechanisms of Hormonal Action in Animals

A. General information

1. Hormones act by altering the metabolism of target cells
   a. These substances can act at extremely low concentrations
   b. A given hormone can affect various target cells differently
2. Hormones may be classified according to two types: steroids and peptides

B. Steroids

1. These hormones exert their effects by entering the target cell's nucleus and influencing the expression of the cell's genes
2. A steroid binds to a receptor protein within the nucleus
3. This hormone-receptor complex binds to an acceptor protein located on the cell's chromatin, thereby activating certain genes

C. Peptides

1. In contrast to steroids, peptides attach to the outer surface of the cell and influence cellular activity through cytoplasmic intermediaries called *second messengers*
2. Peptides bind to specific protein receptors embedded in the plasma membrane; this hormone-receptor complex triggers the release of the second messenger
3. Cyclic adenosine monophosphate and inositol triphosphate are types of second messengers

X. Invertebrate Hormones

A. General information

1. Most invertebrates (animals lacking a spinal column) possess hormones that control water balance
2. The most extensively studied hormones, however, are those that control insect reproduction and growth
3. Insect development is controlled by the interaction of three hormones: ecdysone, brain hormone, and juvenile hormone

B. Ecdysone

1. Ecdysone is secreted from a pair of prothoracic glands just behind the insect's head
2. This hormone triggers the molt, the transition from larva to pupa and from pupa to adult; ecdysone is responsible for turning caterpillars into butterflies

C. Brain hormone

1. This hormone is secreted by the insect's brain
2. It controls the production of ecdysone

D. Juvenile hormone

1. This hormone, secreted by a pair of small glands (called the corpora allata) just behind the brain, balances the action of ecdysone
2. In large quantities, juvenile hormone prevents ecdysone from initiating the molt from pupa to adult

XI. Vertebrate Hormones

A. General information

1. In vertebrates (animals with a spinal column), the internal system of chemical communication is controlled by the endocrine system, which comprises all the tissues and glands that secrete hormones, the hormones themselves, and the molecular receptors on the target cells that respond to the hormones
2. All endocrine glands are ductless and secrete their hormones directly into the circulatory system
3. Hormones differ in the range of their targets: some affect nearly all of the body's tissues, whereas others are more selective, affecting only specific tissues or organs

4. The levels of many hormones are regulated by a negative feedback system, whereby tissues and glands release hormones in response to decreased levels circulating in the bloodstream and stop releasing them when the hormones reach a desired level

B. Hypothalamus

1. This region of the lower brain initiates appropriate endocrine signals after receiving information from other parts of the brain and the peripheral nerves
2. Neurosecretory cells in the hypothalamus release hormones
    a. One set of neurosecretory cells produces hormones that are transported to and stored in the posterior pituitary for later release
    b. A second set of neurosecretory cells produces releasing factors that regulate the actions of the anterior pituitary

C. Pituitary

1. Formerly called the master gland because so many of its hormones regulate other endocrine functions, the pituitary gland is located at the base of the hypothalamus
2. Many pituitary hormones are tropic hormones, which target other endocrine glands
3. The pituitary gland comprises the posterior pituitary and the anterior pituitary
    a. The *posterior pituitary,* also called the neurohypophysis, releases two hormones —oxytocin and antidiuretic hormone (ADH), or vasopressin
        (1) Oxytocin stimulates contraction of uterine muscles and mammary gland cells
        (2) ADH promotes the reabsorption of water from the collecting tubules of the kidneys
    b. The *anterior pituitary,* also called the adenohypophysis, produces seven hormones in response to releasing factors produced by the hypothalamus
        (1) *Growth hormone* stimulates general (but especially skeletal) growth
        (2) *Prolactin* stimulates milk production and secretion in humans
        (3) *Follicle-stimulating hormone* activates the ovarian follicle to secrete estrogens in females and initiates spermatogenesis in males
        (4) *Luteinizing hormone* triggers the corpus luteum to produce progesterone in females and stimulates the interstitial cells of the testes to produce testosterone in males
        (5) *Thyroid-stimulating hormone* prompts the thyroid gland to secrete hormones
        (6) *Adrenocorticotropin* stimulates the adrenal cortex to secrete glucocorticoids
        (7) *Melanocyte-stimulating hormone* regulates the activity of pigment-containing cells in the skin of some vertebrates

D. Thyroid

1. In humans and most mammals, the thyroid gland consists of two lobes located on the ventral surface of the trachea; in other vertebrates, it is located on either side of the pharynx
2. The thyroid gland produces three hormones —triiodothyronine, thyroxine, and calcitonin

a. *Triiodothyronine* (T_3) and *thyroxine* (T_4) target the same organ and mediate the same response, but T_3 usually is more active, achieving the desired physiologic effect but at lower concentrations
   (1) T_3 and T_4 increase oxygen consumption and heat production and maintain metabolic processes; they also play a role in growth and development
   (2) In frogs, T_3 and T_4 control the metamorphosis from tadpole to frog; in humans, a lack of these thyroid hormones at birth results in cretinism, a condition characterized by diminished skeletal growth and mental retardation
b. Calcitonin lowers blood calcium levels

E. Parathyroid

1. This gland is comprised of four tiny glands embedded in the surface of the thyroid gland
2. The parathyroid glands secrete parathyroid hormone, which has the opposite effect of calcitonin —that is, it raises blood calcium levels

F. Pancreas

1. The pancreas is both an endocrine gland (which secretes hormones directly into the bloodstream) and an exocrine gland (which secretes substances into tubes or ducts that empty directly onto the epithelial surface)
2. The exocrine secretions are digestive enzymes, not hormones
3. The three hormones produced by the pancreas are insulin, glucagon, and somatostatin
   a. *Insulin* controls carbohydrate metabolism (by lowering blood glucose), regulates lipid metabolism, and stimulates protein synthesis; it is produced in the pancreas by special beta cells that are part of a group of cells called the islets of Langerhans
   b. *Glucagon* increases blood glucose levels by stimulating glycogen breakdown in the liver; it is produced by special alpha cells, which are also a part of the islets of Langerhans
   c. *Somatostatin* suppresses the release of insulin, glucagon, and growth hormone and is produced by special delta cells that are part of the islets of Langerhans

G. Adrenals

1. Located on top of the kidneys, the adrenals produce hormones in their medulla, or central part of the gland, and their cortex, or outer part of the gland
2. The medulla produces epinephrine and norepinephrine
   a. *Epinephrine* increases blood sugar levels and constricts blood vessels in the skin, mucous membranes, and kidneys
   b. *Norepinephrine* increases the heart rate and force of cardiac muscle contraction and constricts blood vessels throughout the body
3. The cortex produces glucocorticoids and mineralocorticoids
   a. The glucocorticoid *cortisol* increases blood sugar and suppresses parts of the body's immune reaction, such as inflammation
   b. The mineralocorticoid *aldosterone* promotes sodium reabsorption and potassium excretion in the kidneys

H. Gonads

1. These glands produce several hormones, including androgens, estrogens, and progesterone
2. *Androgens* are produced by the testes
   a. Testosterone, the chief androgen, supports spermatogenesis and is responsible for the development and maintenance of male secondary sex characteristics
   b. A small quantity of androgens are produced by the adrenal glands of both men and women
3. *Estrogens* are produced by the ovarian follicle
   a. They initiate buildup of the uterine lining, decrease bone resorption, moderate sodium and water levels in the body, and are responsible for the development and maintenance of female secondary sex characteristics
   b. Estrogens are produced in small quantities by the adrenal glands of both men and women
4. *Progesterone* is produced by the corpus luteum; it promotes growth of the uterine lining

Study Activities

1. Compare negative and positive feedback mechanisms, providing an example of each.
2. Describe the nature and purpose of circadian rhythms.
3. Draw the movement of fluid in animal cells placed in hyperosmotic, hypoosmotic, and isosmotic solutions.
4. Differentiate among conduction, convection, radiation, and evaporation, giving an example of each.
5. Identify the differences in function between steroid and peptide hormones.
6. List the vertebrate hormones that regulate blood glucose levels.

7

Evolution

Objectives

After studying this chapter, the reader should be able to:
- Define microevolution and macroevolution.
- List the six ways in which fossils are formed.
- Describe four lines of evidence that support the theory of evolution.
- List the major assumptions, conclusions, and limitations of Darwin's theory of evolution.
- Identify the five factors involved in the evolution of populations.
- Compare and contrast sympatric speciation with allopatric speciation.
- Relate the geologic eras to evolutionary events.

I. Evidence of Evolution

A. General information

1. Evolution —the processes that transformed all life on earth from its earliest beginnings to its present diversity —provides explanations for the variety, origins, relationships, similarities and differences, geographic distribution, and adaptations of organisms
2. Evolutionary change arises from the interaction of ***populations*** of organisms with their environment
    a. ***Microevolution*** refers to changes in a given *population* (a group of individuals of one species that live in a particular geographic area) over time
        (1) It is concerned with how populations adapt to a changing environment
        (2) It also focuses on how new *species* (organisms that share similar characteristics and can interbreed to produce similar fertile offspring) arise from established species
    b. ***Macroevolution*** refers to the changes of many populations over time; it is concerned with evolutionary trends (such as increasing brain size in mammals) and novel adaptations (such as feathers in birds)
3. A careful examination of several different lines of evolutionary evidence supports two basic themes: organisms progressed from simpler to more complex varieties over time, and all organisms are descended from a common ancestor
4. Scientists have gathered evidence supporting these themes from studies of fossils, comparative anatomy, vestigial organs (useless or marginally useful structures), embryology, classification schemes of living organisms, biogeography, and molecular biology

B. Fossil studies

1. A ***fossil*** is any preserved remnant or impression of a living organism that existed long ago; most fossils are found in sedimentary rock
2. An entire organism, including its soft parts, is rarely fossilized; more commonly, the hard parts of an organism (such as shells, teeth, and bones, which do not decay as rapidly) remain as fossils
3. Fossils can be formed in numerous ways, including petrification; formation of imprints, casts, or molds; freezing; and entrapment in tar pits or amber
   a. *Petrification* occurs when minerals dissolved in ground water seep into the tissues of a dead organism and replace its organic material, thereby turning the plant or animal into stone; the petrified bones of dinosaurs and the petrified wood of trees are formed this way
   b. *Imprints* (such as footprints or leaf prints) are preserved depressions formed in soft mud or sand that subsequently dry and change into sedimentary rock; dinosaur tracks are examples of imprints
   c. *Casts* or *molds* are formed when animals, such as snails, are buried in mud under water for ages; as the organism decays, its shape is replaced by minerals, leaving a cast of the original animal in the surrounding sedimentary rock
   d. *Freezing,* which can occur if an organism is trapped in freezing soil, snow, or ice that does not thaw over time, can prevent bacterial decay and preserve much of the organism; the preserved bodies of prehistoric mammoths have been found in Siberia
   e. *Tar pits* create a sticky material from which animals, once trapped, cannot escape; because minimal bacterial decomposition occurs within these pits, the skeletons of many prehistoric animals, such as saber-toothed tigers, have been preserved
   f. *Amber,* the sticky, hardened resin of evergreen trees, was responsible for trapping many prehistoric insects; because little bacterial decay occurred, the insect bodies have been preserved
4. The age of fossils can be determined by relative dating and absolute dating
   a. *Relative dating* entails noting the layer of sedimentary rock in which the fossil is found and inferring from its position the relative age of the fossil; fossils in the lowest strata must be older than those in higher strata
   b. *Absolute dating* involves the use of radioactive isotopes to determine the ages of rocks and fossils on a scale of absolute time
      (1) Radioactive isotopes that accumulated in the fossil when it was alive can be used to date a specimen because these isotopes have a fixed rate of decay (called the half-life)
      (2) For example, carbon 14 has a half-life of 5,600 years; thus half the carbon 14 in a specimen will be gone in 5,600 years (a specimen with 10 grams of carbon 14 will have 5 grams left after 5,600 years and 2.5 grams left after 11,200 years)
5. Paleontologists (scientists who study fossils) have found evidence indicating that life progressed from simpler to more complex forms
   a. Examination of successive geologic strata (layers) of fossil sites typically reveals that the adjacent higher strata contain not only fossils of organisms found in the earlier layers but also fossils of new and slightly more complex organisms

b. Fossils of the most recent and complex plants and animals are found only in the newest strata

6. Some nearly complete fossil series show the step-by-step progression from simpler to more modern animals; one of the best series is that of the horse
   a. The modern horse (*Equus*) is believed to be a descendant of a much smaller ancestor, *Hyracotherium,* which lived during the Eocene epoch about 40 million years ago
   b. As the horse evolved, its number of toes decreased from four to one on each foot, its teeth were modified for grazing (scraping the ground for herbs and other vegetation) rather than browsing (eating tender shoots and twigs at tree or shrub level), and its size increased
   c. Fossils of horses representing several intermediate stages between *Hyracotherium* and *Equus* have been discovered; the fossil record indicates the following progression: *Hyracotherium* (Eocene epoch), *Mesohippus* (Oligocene epoch), *Merychippus* (Miocene epoch), *Pliohippus* (Pliocene epoch), and *Equus* (Pleistocene epoch)

C. Comparative anatomy

1. Comparative anatomy, the study of anatomic similarities between different species, reveals the presence of homologous structures
2. *Homologous structures* are similar in form but not necessarily in function
   a. Scientists believe that the basic similarity of certain structures may have resulted from a common ancestor
   b. The forelegs, wings, flippers, and arms of different mammals are variations on a common anatomic structure that has been modified for different functions (see *Homologous Structures in Mammals,* page 82)
3. Homologous structures are different from *analogous structures,* which are similar in function but not necessarily in form
   a. Analogous organs evolved independently of each other, not from a common ancestral prototype
   b. Insect wings are analogous to bird wings, but these two structures are not homologous

D. Vestigial organs

1. Vestigial organs are remnants of structures that once served necessary functions but today are no longer useful; their presence indicates descent from common ancestors
2. Human vestigial organs include the appendix, the scalp and ear muscles, the third eyelid (nictitating membrane), and the coccyx (tailbone)

E. Embryology

1. Embryologists, who focus on the stages of development from fertilization to birth, recognize that closely related organisms pass through similar stages as embryos —a finding that supports the conclusion that these organisms evolved from a common ancestor
   a. All vertebrate embryos have gill slits during some stage of their embryonic development

b. In fish, these gill slits develop into gills; in terrestrial vertebrates, they are modified for other functions, such as the eustachian tubes that connect the middle ear with the throat

2. ***Ontogeny***, which focuses on the development of an individual organism, provides clues to an organism's ***phylogeny*** (the evolutionary history of the organism as a member of a genetically linked group); because of early similarities in the development of individual organisms of different phyla, the probability that different phyla emerged from a common ancestor can be taken as proof of descent with modification

F. Classification schemes

1. Taxonomy is the branch of biology charged with naming and classifying the diverse forms of life
2. When plants and animals are classified based on their similarities, a hierarchy of structure (a progression from simple to more complex organisms) becomes apparent; this hierarchy is one of the basic tenets of evolution
3. The first taxonomic classification system, developed by Carolus Linnaeus (1707-1778), remains the basis of modern taxonomy
   a. Linnaeus divided all known forms of life into two kingdoms (Plant and Animal) and classified the various organisms according to a specific hierarchy of increasingly general categories (***species***, genus, family, order, class, phylum, and kingdom)
   b. A *species,* the most specific category within the taxonomy, is a group of organisms whose members share similar characteristics and the capacity to interbreed to produce similar fertile offspring
   c. Linnaeus assigned each species a two-part (binomial) Latin name, according to its generic name (*genus*) and specific name (*species*)
4. In 1969, Robert Whittaker of Cornell University revised the classification system to account for five kingdoms (Monera, Protista, Fungi, Plant, and Animal)
5. Taxonomy is a dynamic discipline that changes with the reexamination and discovery of living species as well as from advances in the fields of molecular biology, embryology, and paleontology; for this reason, all taxonomic boundaries above the species level are considered tentative
6. The modern taxonomic scheme does not account for ***viruses***, which are so different from all living organisms that they must be considered separately (see *Appendix B: Taxonomic Classification of Living Organisms)*

G. Biogeography

1. Biogeography is the study of the geographic distribution of species
2. Islands or other isolated spots typically are home to many species of plants and animals that exist nowhere else in the world; these isolated organisms, however, are closely related to species on the nearest mainland or neighboring island
3. Biogeographic data indicate that the longer animals are apart, the more pronounced the differences that develop between them; eventually, this isolation will produce a new species from the common ancestor

Homologous Structures in Mammals

The homology of structures in the human arm, whale flipper, and bat wing are illustrated here. Although similar in structure, these bones are not necessarily similar in function.

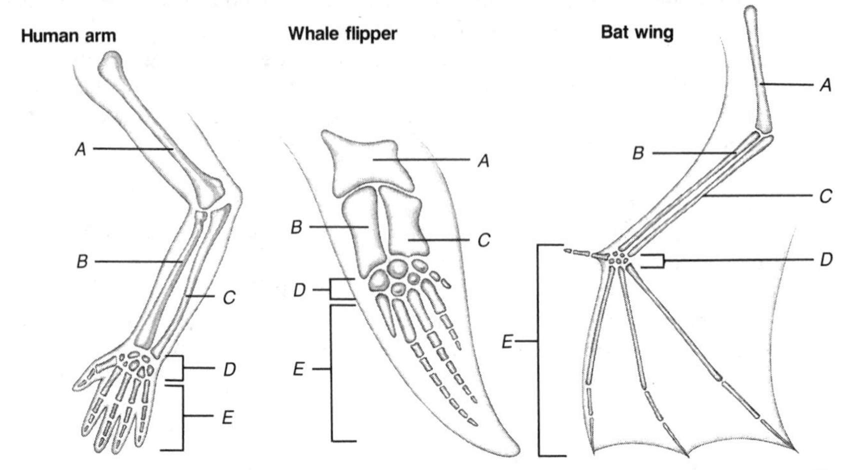

	Human arm	Whale flipper	Bat wing
A (Humerus)			
B (Radius)			
C (Ulna)			
D (Wrist)			
E (Hand)			

H. Molecular biology

1. Molecular biology provides information about the hereditary material (DNA and proteins) that are passed on from one generation to the next
2. The biochemical similarities among different species suggests that all organisms have descended from a common ancestor
3. Scientists have inferred that the closer the similarities in hereditary material, the closer the relationship between the two species
   a. For example, scientists compared the number, types, and position of amino acids in the beta chain of human hemoglobin with that of other species and discovered that the human beta chain differed from the gorilla beta chain by only one amino acid; from the rhesus monkey beta chain by eight amino acids; and the frog beta chain by 67 amino acids
   b. Thus humans are most closely related as a species to apes and, in the example given, least closely related to frogs

II. Pre-Darwinian Theories

A. General information

1. Pre-Darwinian theories are those that predate the work of Charles Darwin and include early ideas about the significance of fossils, the rate of geologic change, and how adaptive traits are inherited
2. The prevailing belief through the end of the 19th century was that species are fixed —that is, they do not change
3. Other early theories established that the earth is extremely old and that slow and subtle processes over time can produce substantial changes; these conclusions greatly influenced Darwin

B. Catastrophism

1. The viewpoint that species are fixed is best exemplified by the work of Georges Cuvier (1769-1832), whose theory is known as *catastrophism*
   a. Cuvier speculated that the boundaries between fossil layers correspond to catastrophic events (such as floods or drought) that destroyed many of the species living at that time
   b. He further postulated that after the extinction of a species in an affected region, foreign species immigrated to repopulate the area
   c. He also maintained that geologic processes were sudden events
2. Cuvier is credited with founding the science of paleontology, or the study of fossils

C. Gradualism

1. In 1795, James Hutton (1726-1797), a geologist, proposed his theory of *gradualism*
2. This theory holds that profound geologic change is the cumulative product of a slow and continuous process, a conclusion opposite that of Cuvier

D. Uniformitarianism

1. Charles Lyell (1797-1875), a geologist, built on Hutton's theory of gradualism and proposed the theory of *uniformitarianism*

2. According to Lyell, geologic processes are so uniform that their rates and effects must balance out through time; in other words, the processes that build mountains eventually are balanced by processes that erode mountains

E. Use and disuse and inheritance of acquired characteristics

1. In 1809, Jean-Baptiste Lamarck (1744-1829) proposed a mechanism to explain how specific adaptations evolve
2. The two principles involved in his theory are use and disuse and the inheritance of acquired characteristics
   a. According to the principle of use and disuse, organs that are used extensively to cope with the environment become larger and stronger, while those that are used infrequently deteriorate; for example, a giraffe stretching its neck in pursuit of new leaves develops a longer neck
   b. According to the principle of acquired characteristics, the modifications that an organism acquires during its lifetime (such as the longer neck of the giraffe) can be passed on to its offspring

III. Darwin's Theory of Evolution

A. General information

1. Charles Darwin (1809-1882) was greatly influenced by the geologic principles of gradualism and applied these principles to biological evolution; these principles influenced him in two important ways
   a. They convinced him that the earth was extremely old
   b. They also convinced him that slow, gradual changes can produce significant results
2. Darwin was the first scientist to explain evolution by maintaining that a new species could evolve from an ancestral form; he argued against the old idea that species are fixed, claiming instead that species evolved over a long period through the gradual accumulation of adaptations to a new environment
3. Darwin was first struck by the geographic distribution of species while serving as a naturalist during a 5-year expedition on a British naval ship; he began to perceive the origin of new species and their adaptations as closely related processes
4. In his account of the expedition (*The Voyage of the Beagle,* 1845), Darwin alluded to his views on evolution and the mechanism by which the overgrowth of organism populations is checked (he later referred to this mechanism as ***natural selection***)
5. Darwin's views were formally presented to the scientific community in 1858 along with the works of Alfred Russel Wallace, a colleague who simultaneously had developed a similar theory on natural selection
6. In 1859, Darwin published *On the Origin of Species by Means of Natural Selection,* which documented his theory of evolution
   a. Darwin's theory of natural selection presented an explanation for how new species arise from established ones
   b. He maintained that a new species originates from an ancestral form by gradually accumulating adaptations to different environments, a process he termed *descent with modification*

B. Major theory assumptions

1. All species have tremendous potential fertility, and their population would increase exponentially if it were not somehow checked
2. Natural resources are limited
3. Individuals in a population display various characteristics, many of which are inheritable

C. Theory conclusions

1. The presence of more individual organisms than the environment can support leads to a struggle for existence, which limits the number of offspring that survive in each generation
2. This struggle for existence is not random; individuals that inherit characteristics best suited to the environment are more likely to produce offspring that will survive
3. This unequal ability to survive and reproduce gradually changes a population over time, with the favorable characteristics becoming more common among all members of the population

D. Theory limitations

1. Darwin recognized that his theory had certain limitations
2. Darwin's theory does not distinguish between inherited and non-inherited variations; Darwin could not explain the mechanical process by which inherited or acquired variations arose in an individual and were passed on to the individual's offspring
3. Darwin realized that natural selection occurs only after individual variations have appeared; however, he could not explain why only natural selection may work on an individual but only populations can evolve

IV. Post-Darwinian Theories

A. General information

1. Post-Darwinian theorists built on Darwin's work
   a. They integrated his principle of natural selection with concepts from the field of population genetics (the study of gene pools and genetic variations in biological populations)
   b. They also modified the Darwinian concept of gradualism
2. Post-Darwinian theorists benefited from knowledge about genetic mutations as well as from a more comprehensive understanding of the fossil record

B. Mutations theory

1. Building on the work of Gregor Mendel and other geneticists, Hugo de Vries (1848-1935) proposed a theory to explain how new characteristics arise in populations
2. He stated that new species result from *mutations,* or sudden changes in their hereditary material
3. Mutations can be beneficial or harmful to an organism
   a. By natural selection, organisms with beneficial mutations survive whereas those with harmful mutations die out

b. New species develop by accumulating useful mutations
c. According to the mutation theory, the modern giraffe arose because a short-necked ancestor produced an offspring with the mutation of a slightly longer neck; this mutation allowed the offspring to feed on the leaves on higher branches, conferring a survival advantage; this beneficial mutation was passed on to succeeding generations

C. The Hardy-Weinberg law

1. In 1908, G.H. Hardy and W. Weinberg proposed a theorum that encompassed many of the basic tenets now associated with *population genetics*—the application of genetics to the theory of natural selection within populations
   a. A *population* is comprised of many individuals, each possessing a unique combination of genes
   b. A *gene pool* comprises the total aggregate of genes in a population at any one time; it consists of all alleles at all gene loci in all individuals of the population
   c. All members of the next generation of a population draw their alleles from the current gene pool
   d. Although individuals can demonstrate genetic changes, only populations can demonstrate the overall prevalence of changes in the gene pool (evolution)
2. The Hardy-Weinberg law states that the frequency of the alleles in a population's gene pool will remain constant from generation to generation if the population is large and no random mating occurs or other new factors (such as mutation or migration) are introduced; because of this genetic constancy, the gene pool of the population is in a state of equilibrium
3. This law describes the genetics of an ideal population that does not interact with other populations and therefore never changes (it is a closed system); the fact that real populations *do* interact and change indicates that evolution (more precisely, microevolution) must occur as a result of some disruption in the genetic equilibrium
4. Various factors—natural selection, mutation, genetic drift, gene flow, and nonrandom mating—can disrupt the Hardy-Weinberg equilibrium, resulting in population changes or microevolution
   a. *Natural selection,* the most common factor affecting genetic variability within a gene pool, is caused by the unequal ability of individuals to produce viable, fertile offspring (a condition of the Hardy-Weinberg law); such differential reproduction results in a decrease in the frequency of some genes and an increase in the frequency of others
      (1) One of the best documented cases of natural selection involves the color changes of the English peppered moth (*Biston betularia*), which over several generations changed primarily from gray to black to gray in response to the environmental effects of pollution and pollution control (see *Natural Selection: The Peppered Moth,* page 88)
      (2) Other examples of natural selection include the prevalence of certain pesticide-resistant insects and antibiotic-resistant bacteria
   b. *Mutation* directly changes the gene pool of a population through substitution of one allele for another

c. *Genetic drift* is a change in the gene pool of a small population from one generation to the next that results from chance; two effects can produce populations small enough to experience genetic drift

(1) In the *bottleneck effect,* a disaster (such as an earthquake, a fire, or a flood) drastically reduces the size of a population, killing its victims unselectively

(2) In the *founder effect,* a few individuals are separated from a larger population, perhaps on an island or in a lake; the smaller, separated population is unlikely to have the same proportion of genes as the larger population

d. *Gene flow* occurs when a population gains or loses alleles from the migration of fertile individuals or the transfer of gametes between populations

(1) Gene flow tends to reduce the differences between populations that occur from genetic drift and natural selection

(2) The ability of humans to move freely about the world is an example of gene flow

e. *Nonrandom mating* occurs when individuals mate with close neighbors, promoting inbreeding that causes the frequencies of genotypes to deviate from those expected by the Hardy-Weinberg equilibrium

D. Punctuated equilibrium theory

1. In the 1980s, Niles Eldridge and Stephen Jay Gould proposed a modification of Darwin's idea of gradualism that combined the knowledge from genetics with a fuller knowledge of the fossil record
2. According to their *punctuated equilibrium theory,* some species remain unchanged for long periods, followed by short periods during which new species are rapidly formed; these short periods are again followed by long periods of relative genetic stability, or equilibrium
3. The fossil record supports this theory; paleontologists have found many examples of new species that seem to appear suddenly but few examples of a slow and steady transformation of species

V. Speciation

A. General information

1. Speciation is the creation of a new species, the largest unit of a population in which gene flow is possible
2. The fossil record documents two types of speciation: anagenesis and cladogenesis

a. ***Anagenesis*** (phyletic evolution) occurs when a species changes so drastically over time that it becomes a different species and the ancestral species ceases to exist; anagenesis results in an unbranched lineage

b. ***Cladogenesis*** occurs when one or more new species develop from an ancestral species, yet the ancestral species continues to exist; cladogenesis results in a branched lineage

c. Cladogenesis is more common than anagenesis and produces greater biological diversity

Natural Selection: The Peppered Moth

The English peppered moth (*Biston betularia*) is a good example of how natural selection can produce population changes. The peppered moth has two phenotypic variations for coloration—a light (gray) color and a dark (black) color. The moth typically feeds at night and rests during the day on trees and rocks covered with light-colored lichens. Moths that stand out against this feeding background are easy prey for birds, and most are eaten before they can pass their genes to the next generation.

Before the Industrial Revolution, black moths were rare because they stood out against the light-colored lichens. However, as pollution from factories killed the lichens and darkened the background against which the moths fed, the white variety became more detectable to predators and the black variety gradually became more prevalent. Today, pollution-control measures have enabled the lichens to reestablish themselves in moth-feeding areas; researchers are beginning to see a shift in the moth population from black to gray.

3. Events that block the gene flow between two populations are critical to the origin of a new species; such barriers may be reproductive or physical in nature
   a. ***Sympatric speciation*** refers to the development of a new species as a result of a reproductive barrier
   b. ***Allopatric speciation*** refers to the development of a new species as a result of a physical barrier

B. Sympatric speciation

1. Sympatric speciation occurs when a subpopulation is reproductively isolated from its parent population
2. This type of speciation usually results from a genetic change that produces a reproductive barrier between the parent population and the mutant offspring
   a. For example, an accident during cell division could result in an extra set of chromosomes, producing a tetraploid —an organism with four complete sets of chromosomes

b. These tetraploids can mate with each other but not with normal diploids, resulting in a new species with a different genetic makeup

3. Sympatric speciation plays a role in plant evolution (for example, polyploid plants are typically more vigorous than plants with the normal number of chromosomes and are likely to survive; however, they can no longer mate within the original plant population and begin to reproduce in isolation, resulting in a new species of plants)
4. Reproductive isolation can result from prezygotic or postzygotic factors
   a. *Prezygotic factors*—ecological, temporal, behavioral, mechanical, or gametic—exert their effects before fertilization
      (1) Ecologic isolation occurs when populations live in different areas and do not interact
      (2) Temporal isolation occurs when mating or flowering takes place at different times of day or during different seasons
      (3) Behavioral isolation occurs when different species are not sexually attracted to each other
      (4) Mechanical isolation occurs when structural differences in genitalia or flowers prevent copulation or pollen transfer
      (5) Gametic isolation occurs when male and female gametes fail to fuse to form a zygote or cannot survive in the female reproductive tract of another species
   b. *Postzygotic factors* (hybrid inviability or sterility) play a role after fertilization occurs
      (1) Hybrid inviability occurs when hybrid zygotes do not develop to sexual maturity
      (2) Hybrid sterility occurs when hybrids do not produce viable gametes; this factor is exemplified by the mule—a sterile hybrid offspring of a horse and a donkey

C. Allopatric speciation

1. Allopatric speciation occurs when a physical barrier geographically segregates a population
2. The finches of the Galapagos Islands illustrate this type of speciation
   a. Thirteen species of finches have evolved on the Galapagos archipelago, all from a single ancestral species
   b. Some of the ancestral population may have been blown from the mainland to another island; this isolated population developed into a separate species, which also may have been blown to a different island, resulting in a third species; this sequence may have been repeated to produce all 13 species
3. The evolution of the Galapagos finches also provides an example of ***adaptive radiation,*** the emergence of numerous species from a common ancestor; in the case of the finches, each species adapted to its new environment by developing a beak specialized for the different types of food on each island

VI. Geologic Timetable

A. General information

1. By dating rocks and fossils, researchers have produced a history of the earth — a timetable that relates geologic eras with evolutionary events (see Appendix A)
2. This geologic timetable illustrates the progression of life from simpler to more complex forms
3. The intervals between eras, periods, and epochs are marked by distinct changes in the species found in sedimentary rocks

B. Precambrian era (roughly 4,600 million years ago)

1. The first type of living cells, prokaryotic organisms, appeared during this era (3,500 million years ago)
2. After the advent of photosynthesis (2,500 million years ago), eukaryotic organisms appeared (1,500 million years ago)
3. Simple multicellular animals (such as sponges and jellyfish) emerged about 700 million years ago, toward the end of the Precambrian period

C. Paleozoic era (600 million years ago)

1. The six periods of the Paleozoic era are the Cambrian (600 million years ago), Ordovician (500 million years ago), Silurian (450 million years ago), Devonian (400 million years ago), Carboniferous (360 million years ago), and Permian (280 million years ago)
2. Invertebrates appeared during the Cambrian period; the first vertebrates (jawless fishes) emerged during the Ordovician period
3. The Silurian period marked the introduction of plants and arthropods and the first colonization of land
4. Amphibians and insects surfaced during the Devonian period, and reptiles and seed plants appeared during the Carboniferous period
5. The Permian period saw an explosion of insect species, including most of the species alive today

D. Mesozoic era (250 million years ago)

1. The Mesozoic era is commonly called the Age of Reptiles because dinosaurs appeared and became extinct during this time
2. This era comprised three periods —Triassic (250 million years ago), Jurassic (215 million years ago), and Cretaceous (144 million years ago)
   a. During the Triassic period, dinosaurs as well as the first mammals and birds appeared
   b. The Jurassic period was dominated by dinosaurs
   c. During the Cretaceous period, flowers first appeared and dinosaurs became extinct

E. Cenozoic era (65 million years ago to present)

1. Also called the Age of Mammals, the Cenozoic era marked the introduction of most modern mammalian species, including humans
2. The two periods of the Cenozoic era are the Paleogene and the Neogene

3. The Paleogene period is further divided into three epochs —Paleocene (65 million years ago), Eocene (54 million years ago), and Oligocene (38 million years ago)
   a. During the Paleocene epoch, a major radiation of mammals, birds, and pollinating insects occurred
   b. Flowering plants (angiosperms) predominated during the Eocene epoch
   c. The Oligocene epoch saw the emergence of apes
4. The four epochs of the Neogene period are the Miocene (24 million years ago), Pliocene (5 million years ago), Pleistocene (1.8 million years ago), and Recent (10,000 years ago to present)
   a. The number of mammal and flowering plant species continued to increase during the Miocene epoch
   b. During the Pliocene epoch, apelike ancestors of humans appeared
   c. Humans emerged during the Pleistocene epoch, which corresponds to the Ice Age

VII. Human Evolution

A. General information

1. The order of primates to which modern humans belong is divided into two subgroups: prosimians (premonkeys) and anthropoids (monkeys, apes, and humans)
   a. The first anthropoids to appear in the fossil record are monkeylike primates that probably evolved from prosimians roughly 40 million years ago in Africa and Asia; modern anthropoids include gibbons, orangutans, chimpanzees, apes, and humans
   b. Ramapithecus is the name given to an anthropoid whose fossils date back about 8 to 14 million years ago; whether Ramapithecus represents the first divergence between humans and the apes or whether it is a common ancestor to both is unclear
2. The evolution of ***hominids*** is marked by the ability to walk erect (bipedalism), an increase in brain size, and the creation of tools
   a. The first hominid, *Australopithecus,* appeared on the African savannas approximately 2 to 3 million years ago; this hominid walked fully erect but its brain was only one-third the size of a modern human's
   b. *Australopithecus afarensis* is thought to have given rise to *Homo habilis,* the first hominid to make and use tools
   c. *Homo erectus* (Peking and Java Man) evolved from *Homo habilis* about 1.5 million years ago; this species, which built fires and resided in huts, stood upright and had a brain capacity of 1,000 ml (compared with modern human's average of about 1,375 ml)
   d. *Homo erectus,* the first hominid to migrate from Africa to Europe and Asia, gave rise to *Homo sapiens*

B. *Homo sapiens*

1. The oldest fossils of *Homo sapiens* are those of the Neanderthals, which lived about 130,000 to 30,000 years ago in the Neander Valley of Germany

2. Neanderthals had heavier brow ridges and less pronounced chins than modern humans but also slightly larger brains; they were skilled tool makers and conducted burials and rituals
3. The oldest fully modern fossils, which date back some 90,000 years ago, belong to another group of *Homo sapiens*—Cro-Magnon man
4. The Neanderthals and Cro-Magnons coexisted, but the Neanderthals eventually became extinct; recent evidence indicates that Neanderthals did not contribute to the modern human gene pool

Study Activities

1. Describe how comparative anatomy supports the theory of evolution.
2. Compare ontogeny and phylogeny.
3. List the taxonomic classification (kingdom through species) for any organism.
4. Outline how pre-Darwinian theories support and contradict Darwin's theory of evolution.
5. Define how the theory of punctuated equilibrium modifies Darwin's work.
6. Identify five factors that contribute to reproductive isolation.
7. Construct a timeline highlighting the major events of the four geologic eras.

8

Ecology

Objectives

After studying this chapter, the reader should be able to:

- List and describe the hierarchical levels of ecologic study.
- Identify the five abiotic factors and their effect on living things.
- Describe the two factors that regulate population fluctuations.
- Give examples of the three different types of symbiotic relationships.
- Define primary and secondary ecologic succession.
- Name the trophic feeding levels that exist in an ecosystem.
- Outline the recycling of nitrogen, carbon, and phosphorus within the ecosystem.
- Describe three terrestrial and two aquatic biomes.

I. Ecologic Hierarchy

A. General information

1. *Ecology* is the scientific study of the interactions among organisms (biotic factors) and between organisms and their environment (abiotic factors)
2. These interactions are organized into hierarchical levels of ecologic study, from smallest to largest
    a. *Populations* are groups of individuals of the same species that inhabit a particular location
    b. *Communities* include all the plant and animal populations of different species that interact in a given environment; communities can be aquatic (pond, river, lake, ocean) or terrestrial (field, desert, cave, forest)
    c. An *ecosystem* describes the relationship of the community to its physical environment
    d. *Biomes* are the earth's major communities classified according to the predominant vegetation of the area that they inhabit; they are characterized by the adaptations of organisms to the particular environment
    e. The *biosphere* is the sum of all the planet's ecosystems; this thin layer of life at the surface of the earth includes the soil, water, and air

B. Focus of ecologic study

1. At the population level, ecologic study focuses on population fluctuations (increases or decreases in population size) and the factors that influence these fluctuations

2. At the community level, the focus is on the interactions among species and the ways in which a community can change over time
3. The primary concern at the ecosystem level is the flow of energy through the system and the cycling of chemicals within the system

II. Abiotic Factors

A. General information

1. The abiotic physical factors present in an organism's environment include light, temperature, water, oxygen, and soil conditions
2. The range of abiotic factors throughout the world has produced many diverse organisms, each with special structures and physiologic mechanisms adapted to particular environmental conditions
3. An organism transplanted to an environment in which the abiotic factors are extremely different from those of its native habitat is unlikely to survive

B. Light

1. All ecosystems are driven by solar energy, although only photosynthetic organisms use this energy source directly
2. In aquatic environments, light is an important limiting factor—plants can live only to the depth that sunlight can penetrate

C. Temperature

1. Temperature affects the metabolism of living things because enzymes necessary for chemical reactions are temperature-sensitive
2. Without extraordinary physical adaptations, organisms are restricted to environments with temperatures ranging from 32° to 122° F (0° to 50° C); some bacteria have enzymes that do not become denatured at high temperatures, enabling the bacteria to continue functioning at temperatures of 194° F (90° C)

D. Water

1. Organisms adapt to the availability and quality of the water in their environment
2. In aquatic environments, the primary problem is maintaining osmolarity; in terrestrial environments, the chief problem is avoiding desiccation (dehydration)
3. Plants can be classified according to their water needs
   a. *Hydrophytes* live in water or marshes; they typically have large, floating leaves and shallow roots
   b. *Xerophytes* live in dry conditions; they have deep root systems, abundant water storage tissue, a thick epidermis, and modified leaves or spines
   c. *Mesophytes* live in environments with average amounts of water; they have a well-developed system of roots, leaves, and stems

E. Oxygen

1. The depth to which terrestrial organisms can live in the soil is determined by the availability of oxygen
2. Aquatic organisms are limited by the oxygen supply in the water

F. Soil conditions

1. The soil's physical composition (clay, sand, rocks, pH level, and mineral content) dictates the types of plants that will grow in it
2. Animals that feed on certain types of plants are restricted to the areas in which these plants grow

III. Populations

A. General information

1. A population is a group of interbreeding organisms that use common resources and are regulated by common abiotic (nonliving) and biotic (living) factors
2. Population ecology is the study of population fluctuations, which are affected by the density and dispersion of the organisms

B. Population density

1. Population density is the number of individuals in a specific area
2. It can be measured by counting all the individuals within a given boundary or by using sampling techniques (sampling involves counting the number of individuals in a representative region of the population and extrapolating the number of individuals within the entire population)

C. Population dispersion

1. Population dispersion is the pattern of spacing of individuals within a particular range (the geographic limit or boundary within which a population lives)
2. Population patterns are classified as clumped, random, or uniform
   a. A *clumped* pattern occurs if the individuals are found in patches throughout the range; for example, plant-eating animals of a particular species may clump together in areas where the plants are plentiful
   b. A *random* pattern (the least common pattern) is seen when the individuals are unpredictably spaced throughout the range (for example, trees are commonly dispersed randomly); random dispersion generally occurs when the individuals of a specific population are not strongly attracted to or repulsed by each other
   c. A *uniform* pattern is characterized by the even spacing of individuals throughout the range (for example, animals may be uniformly spaced because of social behaviors that set up individual territories for breeding, feeding, or nesting)

D. Regulation of populations

1. The size of some populations fluctuates in a regular cycle; however, most populations are regulated by density-dependent and density-independent factors (population control)
2. *Density-dependent factors* intensify as the number of individuals increases; in other words, a greater percentage of the individuals within a population are affected by these factors as the population grows
   a. Limited resources and predation by other species are examples of density-dependent factors

b. Density-dependent factors cause population declines because the death rate increases, the birth rate decreases, or both
c. They can keep the population size fairly constant over time
3. *Density-independent factors* affect the same percentage of individuals in the population regardless of its size
a. Weather and climate are the most significant density-independent factors
b. Density-independent factors can create short-term fluctuations in population size
c. In some populations, density-independent factors may control population size before density-dependent factors become consequential

IV. Communities

A. General information

1. A community has the properties of species richness, equitability, and diversity
a. *Species richness* is the number of species within a given community
b. *Species equitability* is the relative abundance of each species with the community
c. *Species diversity* is a measure of the community's richness and equitability
(1) For example, consider two communities—one made up of 1,000 individuals composing five species, with each species represented by 200 individuals, and the other also made up of 1,000 individuals composing five species, but one species has 600 individuals, and the remaining ones have only 100 individuals each
(2) Both communities are equally rich (five species), but the distribution of species is more equitable in the first community, thus this community is considered more diverse
2. Competition among species within a community plays a role in evolution and affects species diversity
3. Prey-predator relationships and symbiotic relationships are two types of interactions among species in a community

B. Prey-predator relationships

1. Within communities, one species (the predator) may eat another species (the prey)
2. Both plants and animals have developed defenses to protect themselves from predators
3. Plants rely on mechanical and chemical defenses
a. The chemicals produced by a plant can make it distasteful or harmful to predators or can cause abnormal development in insects that eat them
(1) Distasteful chemicals, called secondary compounds, include such diverse substances as strychnine, morphine, nicotine, and tannin
(2) Chemicals that cause abnormal development in insects are analogues of insect hormones, that is, they are structurally similar to these hormones
b. Mechanical plant defenses include hooks or spines
4. Animals defend themselves through behavioral adaptations, camouflage, deceptive markings, chemical defenses, and mimicry; animals with effective physical

or chemical defenses commonly develop bright coloration, called aposematic coloration, to warn predators of their defensive capabilities

a. Behavioral adaptations include running, hiding, or defending against predators

b. Camouflage enables an organism to blend in with its background, making it difficult to spot

c. Deceptive markings that look like eyes or heads may momentarily startle predators, allowing the prey time to escape

d. Chemical defenses, such as the odor produced by skunks or the presence of toxins, can be distasteful or harmful to predators

e. Mimicry, the imitation of one species by another, protects or conceals the prey from the predator

(1) In Batesian mimicry, a palatable species mimics an unpalatable one

(2) In Müllerian mimicry, two unpalatable species mimic each other

C. Symbiotic relationships

1. Symbiosis is a relationship between two different organisms living in direct contact with one another

2. Symbiotic relationships can be of three types: parasitism, commensalism, or mutualism

a. *Parasitism* is a relationship in which one organism derives its nourishment from another and harms the host organism in the process; it is to the ***parasite's*** advantage, however, not to kill its host

(1) A tapeworm, which attaches itself to the intestines of animals and absorbs digested food, is a parasite

(2) Other examples of parasites include bacteria and viruses, which can cause serious illness in their hosts

b. *Commensalism* is a relationship in which one organism benefits from another but does not harm the host organism; barnacles that attach themselves to whales are an example of a commensal relationship

c. *Mutualism* is a relationship in which both organisms benefit from their association; for example, in lichens, algae and fungi coexist, clinging to rocks and trees; the algae provide food for the fungi, and in return the fungi provide water, minerals, and protection from predators

V. Ecologic Succession

A. General information

1. Communities change as the conditions affecting them change

2. The orderly sequence of communities replacing each other in a given order is known as *ecologic succession*

3. Ecologic ***succession*** is determined by various interrelated abiotic factors (including temperature, light, and soil conditions) and biotic factors (including inhibition and facilitation)

a. *Inhibition* results when the presence of one or more species deters the presence of other species; for example, mature oak trees shade the ground, preventing other trees from growing

b. *Facilitation* occurs when the presence of one or more species encourages the presence of other species; for example, alders lose their leaves and decompose on the forest floor, lowering the soil's pH; this facilitates the growth of spruce trees, which thrive on acidic soil

4. In a typical ecologic succession, one group of organisms representing one successional stage alters the abiotic factors in a manner that paves the way for species in the next successional stage; for example, the decomposing leaves of one plant species may lower the pH level of the soil, creating an ideal situation for another species that requires acidic soil
5. Ecologic successions are classified as primary or secondary; the final community in the ecologic succession is called the *climax community*

B. Primary succession

1. Primary succession is the population of areas previously barren of life
2. This situation can occur whenever land is laid bare, such as when a new island is formed by a volcano, when strip-mined land is reclaimed, or when a glacier begins to retreat
3. Primary succession is illustrated by the sequence of events that occurred after glaciers retreated in the northwestern United States: The barren landscape gave rise to moss and lichens, which led to alders and cottonwoods, and finally to spruce and hemlock; this succession was completed within 200 years

C. Secondary succession

1. Secondary succession is the return of an area to its natural state when an existing community is cleared away by a disturbance that leaves the soil intact
2. Secondary succession is illustrated by the sequence of events that takes place in the mid-Atlantic United States after a farm field is abandoned: The herbaceous community is supplanted by shrubbery, then pines, and ultimately by oaks and hickories; this process is completed in about 200 years

D. Climax community

1. The permanent or final community that results from ecologic succession is called the climax community
2. The various living things within the climax community exist in equilibrium, and the community can continue indefinitely if left undisturbed
3. The final community takes its name from the dominant types of plants that characterize it; for example, the maple beech climax community (found in the northeastern United States) or the beech magnolia climax community (found in the southeastern United States)

VI. Ecosystems

A. General information

1. An ecosystem comprises the interactions between a community and its environment; two of the most interesting interactions involve the flow of energy and the cycling of chemical elements

2. An organism's *niche* within an ecosystem is the sum total of its use of the biotic and abiotic resources of the environment; two species that occupy the same niche will compete for food and reproduction sites

B. Energy flow

1. The flow of energy within an ecosystem is determined by a series of *trophic* (feeding) levels
   a. The first ***trophic level*** comprises autotrophs (organisms that use light energy to make their own food); such organisms are considered *primary producers* and support all the other feeding levels in the ecosystem
   b. Subsequent trophic levels consist of ***heterotrophs*** (organisms that directly or indirectly depend on the photosynthetic output of autotrophs for energy); such organisms are considered *consumers* and may occupy three or more trophic levels in the ecosystem
      (1) ***Primary consumers*** (the trophic level just above autotrophs) consist of herbivores, or organisms that eat plants or algae (such as insects, snails, and seed-eating birds)
      (2) ***Secondary consumers*** consist of carnivores, or organisms that eat herbivores (such as spiders, frogs, insect-eating birds, and mammals that consume grazing mammals)
      (3) *Tertiary consumers* consist of carnivores that eat other carnivores (such as snakes and owls)
   c. Some organisms, such as decomposers and omnivores, derive their energy from many trophic levels
      (1) *Decomposers* are organisms, such as bacteria and fungi, that feed on dead organisms and waste products from living animals
      (2) ***Omnivores*** are organisms, such as humans, that eat both plants and animals
2. The transfer of energy from one trophic level to another is commonly depicted as a ***food chain;*** however, the complex interrelationship between trophic levels in most ecosystems is more appropriately described as a ***food web***
3. The amount of energy available within an ecosystem is a function of its overall *productivity,* which in turn determines the total amount of organic material (***biomass***) in the ecosystem
   a. *Primary productivity* is the rate at which light energy is converted to chemical energy (organic compounds) by autotrophs
      (1) The ecosystem's biomass depends on the primary productivity of the autotrophs
      (2) At some point, the autotrophs' productivity is slowed by the limited availability of specific nutrients (such as nitrogen or phosphorus); then, the productivity either remains the same or decreases until the nutrient is replenished
   b. *Secondary productivity* is the rate at which heterotrophs incorporate organic material into new biomass
4. At each successive trophic level, the amount of energy decreases, resulting in a decrease in the amount of biomass that can be supported at each level; this loss of energy can be represented graphically by a food pyramid (see *Food Pyramid,* page 101)

C. Cycling of chemical elements

1. Because ecosystems have a limited supply of chemical elements, the continuation of life depends on the recycling of these elements, particularly nitrogen, carbon, and phosphorus
2. Nitrogen, which constitutes about 80% of the earth's atmosphere, is needed by plants and animals to synthesize amino acids, proteins, and nucleic acids; however, before it can be used by living organisms, atmospheric nitrogen (N_2) must be *fixed*—that is, converted into ammonia (NH_3)
3. The *nitrogen cycle,* the process by which nitrogen is recycled in the ecosystem, involves several steps: ***nitrogen fixation***, ammonification, nitrification, assimilation, and denitrification; bacteria play an important role throughout the nitrogen cycle (see *The Nitrogen Cycle,* page 102)
   a. *Nitrogen fixation,* the process by which N_2 is converted (fixed) to NH_3, is carried out by two groups of bacteria: those that live in the roots of legumes and those that live directly in soil or water
      (1) The first group of nitrogen-fixing bacteria (*Rhizobium*) reside in nodules within the root systems of legumes (plants belonging to the pea family); these bacteria have a symbiotic relationship with the plant, releasing nitrogen to the plant in a form suitable for the synthesis of amino acids and in return receiving carbohydrates and other necessary organic compounds
      (2) The second group (primarily blue-green algae) live freely in the soil and water; these bacteria release NH_3 directly into the surrounding area, where it is oxidized by nitrifying bacteria
   b. *Ammonification* is the process by which decomposer bacteria and certain fungi convert organic nitrogen to ammonia; dead plants and animals are the sources of organic nitrogen
   c. *Nitrification* is the process by which NH_3 is oxidized to nitrite (NO_2) and then to nitrate (NO_3)—the nitrogen source preferred by most higher plants; this process is carried out by nitrifying bacteria
   d. *Assimilation* is the absorption of nitrates by plants for use in the synthesis of organic compounds; animals cannot directly use nitrates released by bacteria and must obtain their nitrogen by eating plants or other animals
   e. *Denitrification* is the process by which unused nitrates are converted back into N_2
4. Carbon—essential to all organic compounds—is recycled through a reciprocal process involving respiration and photosynthesis known as the *carbon cycle* (see *The Carbon Cycle,* page 103)
   a. Carbon dioxide and water are released through respiration, the decomposition of plants and animals, and the burning of fossil fuels
   b. Plants take in carbon dioxide and water during photosynthesis and release oxygen in return
5. Phosphorus, another essential element needed by all living organisms, is released into the soil from weathering rocks in the form of phosphate (PO_4); the process by which this element is recycled in the ecosystem is called the *phosphorus cycle* (see *The Phosphorus Cycle,* page 103)
   a. Plants can absorb PO_4 and use it for organic synthesis in various nucleotides, such as ATP; animals obtain PO_4 by eating plants

Food Pyramid

Autotrophs (represented by the unshaded area) occupy the first trophic level of the food pyramid. The primary producers in the ecosystem, these organisms are highly productive in converting light energy to chemical energy and constitute a large percentage of the total biomass. Heterotrophs (represented by the shaded area) are progressively less productive in converting organic material into new biomass; these consumers compose a smaller percentage of the total biomass with each successive trophic level.

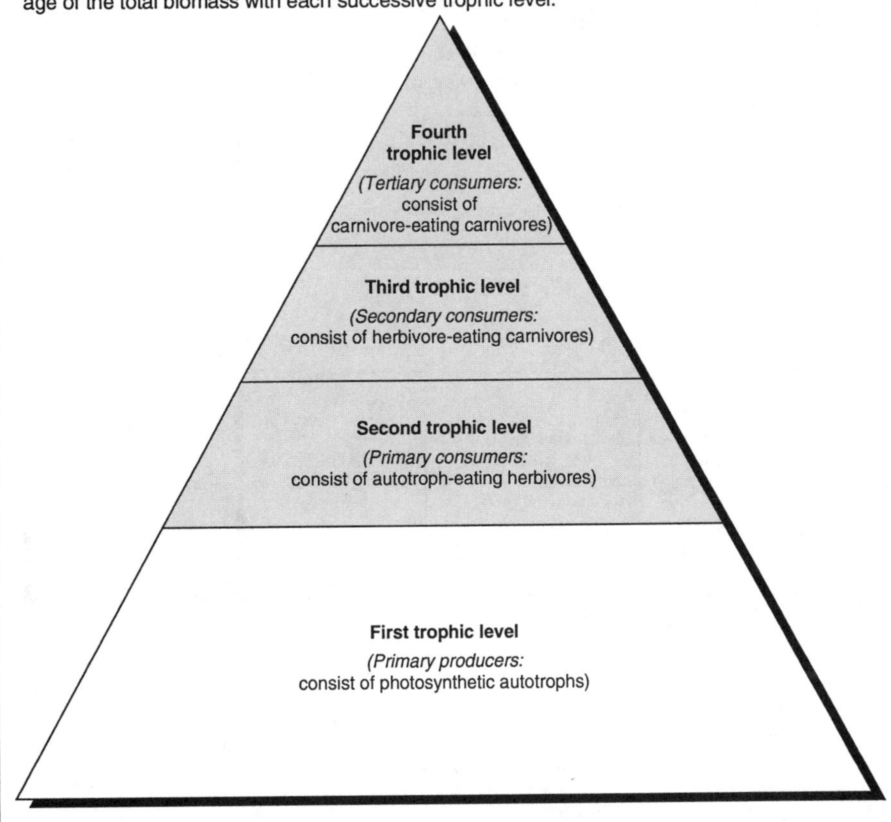

b. Animal excretion and the decomposition of both plants and animals releases phosphate back into the soil; much of the phosphate in soil is washed away into the ocean, where it eventually becomes incorporated into sedimentous rock

VII. Biomes

A. General information

1. A *biome* is an aggregate of climax communities within a particular region that is determined largely by climate
2. Biomes can be classified as terrestrial or aquatic; precipitation and temperature account for much of the variation in terrestrial biomes

The Nitrogen Cycle

All living organisms require nitrogen to synthesize organic compounds. Although nearly 80% of the earth's atmosphere consists of nitrogen, plants and animals cannot use nitrogen in this gaseous form (N_2). Plants can only incorporate nitrogen in the form of ammonia (NH_3) or nitrate (NO_3), and animals can assimilate nitrogen only by eating plants. Eventually, all nitrogen is released from organisms and recycled back into the atmosphere. The overall process of nitrogen cycling is accomplished in five steps: nitrogen fixation, ammonification, nitrification, assimilation, and denitrification.

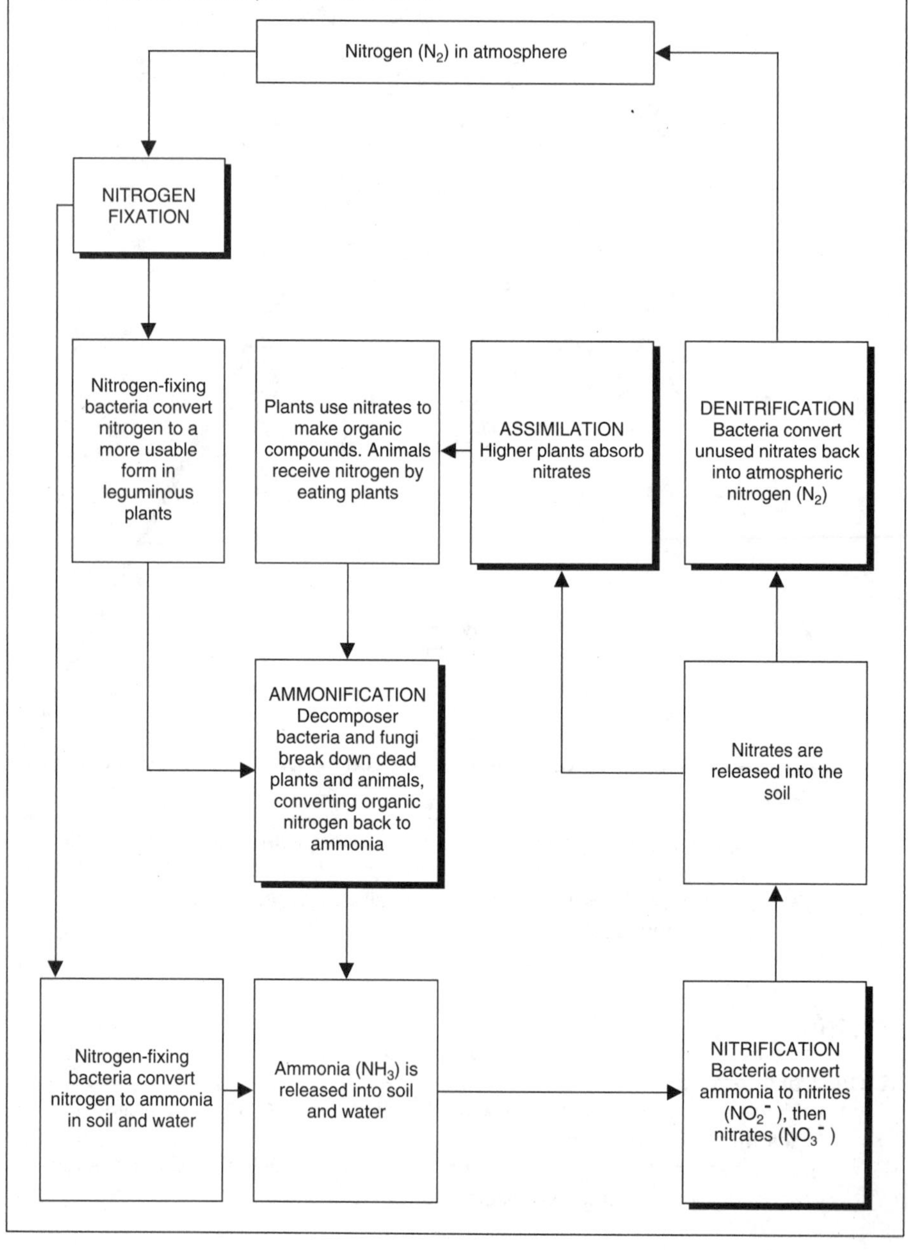

3. The major biomes are classified as tropical rain forests, savannas, deserts, chaparrals, temperate grasslands, temperate forests, taigas, tundras, lakes and ponds, rivers and streams, and marine communities

The Carbon Cycle

The carbon cycle is the process by which carbon —in the form of carbon dioxide (CO_2) — is extracted from and returned to the atmosphere by living organisms. This is accomplished primarily through the reciprocal processes of photosynthesis and respiration, as well as by the decay of plants and animals and the burning of fossil fuels.

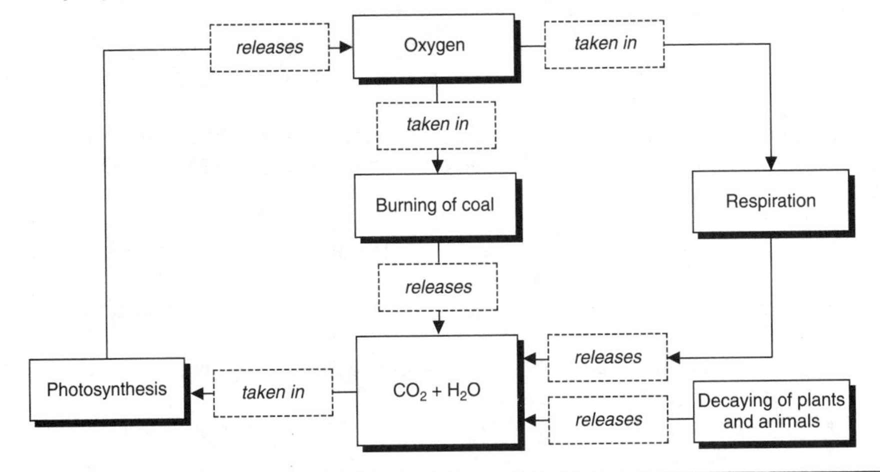

The Phosphorus Cycle

Phosphorus is continuously recycled in the ecosystem because it exists in such limited amounts. The phosphorus cycle begins with the slow release of phosphorus — in the form of phosphate (PO_4) — into the soil by weathering rocks. Plants then absorb PO_4 from the soil to help synthesize organic compounds. Animals, in turn, eat plants and excrete PO_4 back into the soil. Some residual PO_4 also is released when plants and animals decompose.

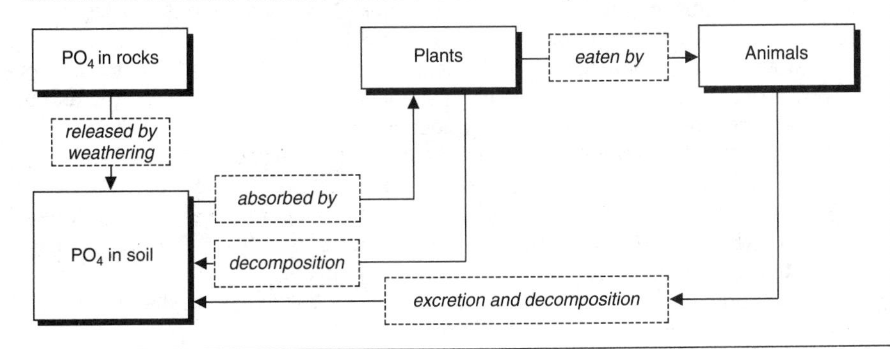

B. Tropical rain forests

1. These biomes, found in areas near the equator where rainfall is abundant and the dry season is short, are the most complex of all communities; rain forests contain more species of plants and animals than any other community in the world
2. Competition for light is keen among plants because the dense trees create an overhead canopy that prevents light from reaching the ground; when an opening in the canopy occurs, many plant species compete to fill the gap
3. Many rain forest trees are covered with epiphytes (plants that nourish themselves but grow on the surface of another plant, usually for support), such as orchids and bromeliads
4. Poor soil is characteristic of this biome because high temperatures and significant rainfall cause rapid decomposition and recycling and little buildup of organic material
5. Many plants depend on mutual relationships with animals to deliver pollen
6. Tree-dwelling and ectothermic animals are typical rain forest inhabitants

C. Savannas

1. These grasslands with scattered individual trees are found in central South America, central and south Africa, and Australia
2. A typical savanna has three distinct seasons —cool and dry, hot and dry, and warm and wet
3. Some of the world's largest herbivores (giraffes, zebras, antelopes, buffalo, and kangaroos) populate savannas
4. Burrowing animals and nocturnal animals are common (few trees are available for protection and nesting sites)
5. These biomes encourage rapid plant growth, which provides an abundant food source for animals

D. Deserts

1. A desert is an area of low precipitation but not necessarily high temperature
   a. Hot deserts are found in the southwestern United States, the west coast of South America, north Africa, the Middle East, and central Australia
   b. Cold deserts are found west of the Rocky Mountains, in eastern Argentina, and throughout much of central Asia
2. Growth and reproduction cycles in deserts coincide with rainfall, not temperature
3. Arid deserts have no perennial vegetation; in less arid deserts, widely scattered shrubs and cacti or succulents predominate
4. Desert-dwelling animals include seedeaters (such as ants, birds, and rodents) and reptiles (such as lizards and snakes) that feed on the seedeaters
5. Most animals are adapted to conditions of high temperature and scarce water
   a. Some animals live in burrows and are active at night
   b. Other animals are light-colored to reflect the sunlight
   c. Many animals have anatomic and physiologic adaptations that allow them to conserve water

E. Chaparrals

1. These mid-altitude areas along coasts are exposed to cool offshore ocean currents; their climate is characterized by mild, rainy winters and long, hot, dry summers
2. Chaparrals are found along the coastlines of California, Chile, southwestern Africa, and southwestern Australia
3. Dense, spiny shrubs with tough evergreen leaves are typically found in this biome; annual plants are abundant during the winter and early spring when rainfall is plentiful
4. Vegetation is adapted to the frequent fires that occur in chaparrals
   a. Plant root systems quickly regenerate and use the nutrients released by fire
   b. Many chaparral species produce seeds that germinate only after a fire
5. Animals found in the chaparral include deer, fruit-eating birds, rodents that eat the seeds of annual plants, and some types of lizards and snakes

F. Temperate grasslands

1. Similar to tropical savannas, this type of biome is found in regions with colder temperatures, such as the plains and prairies of the United States, the steppes of Russia, the pampas of Argentina, and the veldts of South Africa
2. Occasional fires and drought prevent woody shrubs and trees from becoming established, thereby preserving the grasslands characteristics of these areas
3. Large grazers, such as bison, gazelles, and zebras, and large carnivores that feed on these grazers, live in the temperate grasslands

G. Temperate forests

1. Found in most of the eastern United States, middle Europe, and parts of eastern Asia, temperate forests grow in mid-altitude regions where the moisture is adequate to promote the growth of large, broad-leaf, deciduous trees
2. Cold winters and hot summers, with rainfall almost evenly distributed throughout the year, characterize this biome
3. Several layers of vegetation, including herbs and shrubs, are found in these regions; the dominant tree species may be oak, birch, hickory, beech, or maple
4. The variety and abundance of food found in temperate forests support diverse animal species

H. Taigas

1. Characterized by coniferous or boreal forests, taigas are found across North America, Europe, Asia, and the southern border of the arctic tundra, as well as in higher elevations in more temperate latitudes (for example, the mountainous regions of the western United States)
2. This biome is characterized by harsh winters and short summers, with considerable precipitation, mostly in the form of snow
3. The thin, acidic soil supports conifers (such as spruce, pine, fir, and hemlock), which grow in dense stands, as well as some deciduous species (such as oak, birch, willow, alder, and aspen)
4. The snow, which can accumulate to several meters each winter, helps insulate the ground and permits mice and other small mammals to remain active in snow tunnels

5. The animals that populate the taiga are mostly seedeaters (squirrels and jays), browsers (deer, moose, elk, snowshoe hares, beavers, and porcupines), and carnivores (bears, wolves, lynx, and wolverines)

I. Tundras

1. This type of biome is found in the coldest parts of the world, either the northernmost regions (the arctic tundra) or in high-altitude areas (alpine tundras)
2. The arctic tundra encircles the North Pole southward to the taigas
   a. In the arctic tundra, the ground is continuously frozen —a condition that creates permafrost
   b. Permafrost prevents the roots of plants from penetrating deeply into the soil, which limits the types of plants found, especially taller trees
   c. Dwarf perennial shrubs, sedges, grasses, mosses, and lichens predominate; these plants have bursts of productivity and reproductivity during the brief summers, when 24 hours of continuous daylight is normal
3. Alpine tundras are found above the treeline of mountains
   a. They even can be found in the tropics if the elevation is high enough
   b. The amount of daylight in these areas remains constant, at about 12 hours per day, which encourages the growth of vegetation throughout the year
4. Tundra animals are adapted to cold weather; they typically live in burrows or migrate to avoid the cold
   a. Ectothermic animals are rare (except for gnats and mosquitoes in the summer)
   b. Large animals include herbivores, such as musk oxen, caribou, and reindeer; smaller animals include lemmings and a few predators, such as the white fox and snowy owl

J. Lakes and ponds

1. These standing bodies of water range in size from a few square meters to thousands of kilometers
2. The presence or absence of light creates two major zones —the photic zone (where light penetrates) and the aphotic zone (where light cannot reach)
3. In the *photic zone,* sufficient light is available for photosynthesis, which encourages the growth of various ***phytoplankton*** (algae and Cyanobacteria); zooplankton, mostly rotifers and small crustaceans, graze on the phytoplankton; zooplankton are eaten by many small fish, which in turn become food for larger fishes, semiaquatic snakes, turtles, and fish-eating birds
   a. Phytoplankton growth is limited by the amount of nitrogen and phosphorus present in the biome
      (1) A large influx of these nutrients (such as runoff from agricultural fields or fertilized lawns) can lead to an algal bloom, or a population explosion
      (2) An algal bloom can have serious implications for the other lake or pond communities because it quickly depletes the available nutrients
   b. After dying, the smaller organisms of the photic zone sink into the deeper areas, creating a continuous rain of dead organic material, or detritus, into the aphotic zone
4. In the *aphotic zone,* microbes and other organisms feed on the detritus from the photic zone

5. If no mixing occurred between the two zones, life would become extremely limited in both; however, this is not the case
    a. As winter approaches, the surface water cools, becomes denser, and sinks to the bottom, which mixes the deeper water with the surface waters
    b. This mixing restores oxygen (from the atmosphere) to the deeper layers and brings rich nutrients (from the decomposing photic organisms) to the surface
6. In shallower areas, a third zone (***benthic zone***) is evident; aquatic plants, diverse attached algae, clams, snails, crustaceans, and insects inhabit this zone

K. Rivers and streams

1. These bodies of water move continuously in one direction
2. The biological communities of rivers and streams change from the source (headwaters) to the point at which the rivers or streams empty into a lake or an ocean
    a. At the source, the water usually is cold and narrowly channeled, with few nutrients and a rocky substrate; cold-water fish, such as trout, may be found here
    b. The water downstream is warmer, turbid, and more widely channeled, with more nutrients; catfish and carp may inhabit this area
3. The flowing waters of rivers and streams prevent plankton growth; the food chain is supported by the photosynthesis of attached alga and the influx of organic material carried into the stream

L. Marine communities

1. The marine environment can be divided into several zones, each with its own characteristic communities
2. Photic and aphotic zones are defined on the basis of light penetration
3. Intertidal, neritic, and oceanic zones are defined on the basis of distance from the shore and water depth; open water is called the *pelagic zone* and the bottom of the ocean (seafloor) is the *benthic zone*; the deepest benthic area is the *abyssal zone*
4. The *intertidal zone* is subject to daily tides; at one point during the day it is wet, at another point it is dry; the nature of the intertidal community is determined by the type of substrate (rock, sand, or mud)
    a. Rocky intertidal areas are inhabited by barnacles, mussels, sponges, echinoderms, algae, and other organisms that can attach to the rocks
    b. Sandy intertidal zones (beaches) or mudflats are home to organisms that can filter food out of water brought in by waves, such as clams and worms
5. The *neritic zone* lies close to the shore; here, sunlight can penetrate to the bottom of the ocean floor, and many varieties of algae, plankton, and fishes can be found
6. The *oceanic zone* is the farthest from the shore; because of the depth, sunlight cannot penetrate to the bottom of the ocean floor
    a. Algae in this zone are typically flat and contain special air bladders that enable them to float
    b. Plankton undergo a diurnal migration, sinking to the lower depths during the day and surfacing at night
    c. Fish and marine mammals follow this diurnal migration; hence, the surface waters are virtually devoid of life during the day but teem with life at night

7. The benthic community consists of fungi, bacteria, sponges, burrowing worms, sea stars, crustaceans, fish, sea anemones, and clams
8. The area where fresh water meets salt water is called an *estuary;* oysters, crabs, and many fish species inhabit these regions

---

Study Activities

1. Compare a community to a biome.
2. Classify plants according to their water needs and provide an example of each class.
3. Define the three patterns of population dispersion.
4. Identify three methods of defense in prey-predator relationships and give an example of each.
5. Compare a climax community to a community undergoing succession.
6. Describe what happens to productivity at each successive trophic level.
7. Describe the interaction between photosynthesis and respiration in the carbon cycle.
8. Make a representative drawing of a marine community and label each zone.

9

Viruses, Monera, and Protista

Objectives

After studying this chapter, the reader should be able to:

- Describe the structure of a virus and the methods by which a virus reproduces.
- Outline the basic differences among bacterial, plant, and animal viruses.
- List the characteristics of six types of eubacteria.
- Describe three types of protozoa, algal protists, and fungus-like protists.

I. Viruses

A. General information

1. A *virus* is a particle (***virion***) of a nucleic acid (deoxyribonucleic acid [DNA] or ribonucleic acid [RNA]) surrounded by a protein coat
2. A virus is extremely small —in some cases, only about 1⁄50 of a micron in diameter, which is smaller than a ribosome
   a. About 1,000 average-size viruses could fit inside an average-size bacterium
   b. The largest virus barely can be seen with the light microscope
3. A virus can carry out basic life processes (such as reproduction) only when infecting a host organism; for this reason, a virus is not commonly considered a living organism but rather an inert particle with parasitic properties
4. Because the structure and reproductive methods of viruses differ greatly from that of all other organisms, viruses are treated as a separate category within the taxonomy classification system
5. Each virus exhibits host specificity —that is, it can infect only a limited range of host cells; based on this specificity, scientists have identified three categories of viruses: bacterial, plant, and animal

B. Viral structure

1. The protein coat that encloses the viral genome ***(capsid)*** is composed of individual protein subunits called *capsomeres*
2. A capsid may be rod-shaped, polyhedral, or a more complex configuration
3. Some viruses contain accessory structures that facilitate their infection of host cells
   a. Many animal-infecting viruses have a membranous *envelope*—a mixture of phospholipids and proteins derived from the host and proteins and glycoproteins derived from the virus —surrounding the capsid; this structure helps the virus to fuse with the host's cell membrane

b. Many bacteria-infecting viruses have a flexible, rod-shaped tail consisting of proteins; this helps the virus to attach to bacteria

C. Viral reproduction

1. As obligate intracellular parasites, viruses must infect a living host cell to reproduce
2. Once inside a host cell, the virus uses the cell's enzymes, ribosomes, nutrients, and other resources to replicate its genome and produce a protein coat
3. A virus can produce hundreds of thousands of progeny in each generation; each viral offspring leaves the host cell to infect another host cell
4. The viral ***genome*** is composed of either DNA or RNA
   a. Viruses whose genomes consist of single- or double-stranded DNA replicate in a fashion similar to that of nucleic acids within cells; this pattern is described as DNA → DNA
   b. Viruses whose genomes consist of single- or double-stranded RNA rely on special enzymes —RNA replicase or reverse transcriptase —for replication
      (1) RNA replicase enables viruses to use RNA as a template to make complementary RNA, a pattern described as RNA → RNA
      (2) Reverse transcriptase enables viruses to use RNA as a template to make DNA, which is used to transcribe messenger RNA and genomic RNA; this pattern is described as RNA → DNA → RNA

D. Bacterial viruses

1. Most bacteria-infecting viruses, known as ***bacteriophages,*** have a viral genome consisting of single- or double-stranded DNA
2. *Virulent bacteriophages* must kill their hosts to replicate; *temperate bacteriophages* can reproduce with or without killing their hosts
3. The reproductive pattern of a virus that kills its host is termed a ***lytic cycle;*** that of a virus that does not kill its host is called a ***lysogenic cycle***
   a. The lytic cycle involves a series of sequential steps that can be completed in approximately 20 to 30 minutes
      (1) The bacteriophage first attaches itself to the cell surface of the bacterium, then injects its DNA into the cell (the bacteriophage's protein coat remains outside the bacterial cell)
      (2) The bacterial cell begins to transcribe the viral genome; one of the first viral genes transcribed is for an enzyme that hydrolyzes the bacterial cell's own DNA
      (3) Controlled by the bacteriophage's DNA, the host cell manufactures large quantities of the viral DNA and protein coats, which are assembled into mature bacteriophages
      (4) As a final step, the bacteriophage's genome transcribes an enzyme that digests the bacterial cell wall; hundreds of bacteriophages are released, each capable of infecting another cell
   b. In the lysogenic cycle, the bacteriophage inserts its genome into the bacterial genome; after insertion, the viral genome is referred to as a ***prophage***
      (1) The prophage is inactive except for one gene that codes for a repressor protein, whose function is to keep all other prophage genes switched off

(2) When the bacterial cell replicates, the prophage also replicates and passes its genes onto to the two daughter cells; in this way, as the population of bacteria develops, each bacterium contains a copy of the original prophage

(3) Certain environmental factors, such as radiation or chemical exposure, may initiate a process in which the prophage leaves the bacterial genome; these liberated bacteriophages can enter other bacterial cells and begin a lytic or lysogenic cycle all over again

4. To defend against bacteriophages, bacteria have developed bacterial mutations (which can change the receptor sites on the surface of the bacterium, making attachment difficult for the bacteriophage) and special enzymes (which can attack and break down the viral DNA before it takes control of the cell)

E. Plant viruses

1. Most plant viruses have rod-shaped capsids and a genome composed of RNA
2. They are transmitted horizontally or vertically
   a. In *horizontal transmission,* the plant is exposed to an outside source that causes the infection, such as the wind, contaminated pruning tools, or insects that carry the virus from plant to plant
   b. In *vertical transmission,* the plant inherits the virus from its parent; the virus can be passed on by cuttings (asexual reproduction) or through an infected seed (sexual reproduction)
3. Once a virus particle is inside a plant cell and begins to reproduce, it can spread throughout the plant by passing through plasmodesmata (cytoplasmic structures that connect and penetrate the cell walls of adjacent plant cells)

F. Animal viruses

1. Animal viruses cause a wide range of maladies in all animals; for example, rabies, distemper, and numerous types of tumors found in various domestic and wild animals are caused by viruses; in humans, respiratory disease, influenza, cold sores, chicken pox, measles, mumps, polio, diarrhea, some types of cancer, and acquired immunodeficiency syndrome (AIDS) are but a few of the diseases caused by viruses
2. In addition to their capsid, many animal viruses have a membranous envelope, which enables the virus to fuse with the host's cell membrane as a means of replication
3. In a ***productive cycle*** of replication, the virus leaves the infected cell by budding, a process that does not necessarily destroy the cell
   a. The productive cycle of replication is characteristic of all animal cells
   b. During this cycle, the virus fuses with the cell membrane of the host cell and injects its own DNA into the host cell
   c. Once integrated into the host's genome, the virus is called a ***provirus***
4. A ***retrovirus*** is a type of virus that carries the enzyme ***reverse transcriptase***, which catalyzes the transcription of RNA into DNA; human immunodeficiency virus (HIV), the virus that causes AIDS, is an example of a retrovirus
   a. The envelope of HIV contains knoblike proteins (gp 120) that enable the virus to bind to specific receptor sites on the surface of the host cell
   b. HIV infects a subpopulation of human white blood cells called T_4, or helper, cells; these cells are an integral part of the body's immune response

c. A person whose T_4 cells are infected with HIV cannot maintain an effective immune response against invading pathogens, which can lead to the opportunistic infections characteristic of AIDS, such as *Pneumocystis carinii* pneumonia, tuberculosis, esophageal candidiasis, and gastroenteritis
d. HIV is transmitted via infected blood, semen, or body fluids; once this virus invades the bloodstream, it attaches to and fuses with a T_4 cell, then injects its genome and core proteins into the cell (see *The Productive Cycle of HIV)*
e. The proteins remain with the RNA as it is copied into DNA; the original RNA is then destroyed and the duplicated DNA is incorporated into the T_4 cell's own DNA (which is now called a provirus)
f. The provirus can remain latent and cause no changes in the cell, or it can be activated and take control of the cellular mechanisms of the T_4 cell
g. In the activated phase, the viral RNA makes proteins, assembles itself into a new HIV particle, and buds from the cell
h. During budding, the T_4 cell may be destroyed or spared
i. The new HIV particle enters the bloodstream to infect other T_4 cells

5. Because viruses have few or no enzymes of their own, antibiotics (which work by inhibiting enzymes or specific biosynthetic processes) are ineffective in treating viral illnesses

II. Monera

A. General information

1. All organisms in the kingdom Monera are prokaryotic cells, the earliest known forms of life on earth
2. Most are unicellular, although some species consist of aggregates of cells that stick together after dividing
3. All monera reproduce asexually by means of binary fission; no mitosis or meiosis takes place, although occasionally recombination —the transfer of genes between chromosomes —may occur
4. Monera obtain nutrition through various methods
   a. *Photoautotrophs* harness light energy to synthesize organic compounds from carbon dioxide (photosynthesis)
   b. *Photoheterotrophs* use light energy to generate adenosine triphosphate but obtain carbon in organic form through absorptive feeding
   c. *Chemoautotrophs* use energy from the oxidation of inorganic substances (chemosynthesis) and need only carbon dioxide as a carbon source
   d. *Chemoheterotrophs* must consume organic molecules for energy and to obtain carbon
5. Oxygen has a variable effect on monera
   a. *Obligate aerobes* use oxygen for cellular respiration and cannot grow without it
   b. *Facultative anaerobes* use oxygen when available but also can grow by fermentation in an anaerobic environment
   c. *Obligate anaerobes* cannot use oxygen because they are poisoned by it
6. Classifying the more than 10,000 known species of prokaryotic organisms is difficult; at present, no definitive phyla classifications are available

The Productive Cycle of HIV

During the productive cycle of replication, human immunodeficiency virus (HIV) fuses with its host cell (a human T_4 cell, which is essential to the proper functioning of the immune system) and injects its genome into the cell. Incorporation of the viral DNA into the T_4 cell's DNA results in a provirus, which may reproduce and form new HIV particles that exit the T_4 cell by budding.

HIV structure

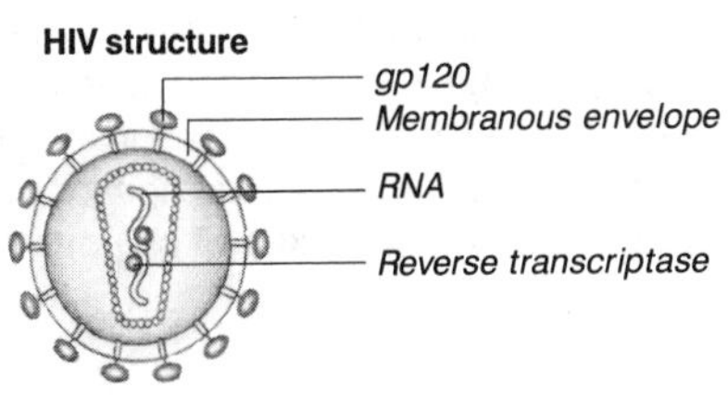

Productive cycle

HIV fuses with T_4 cell wall, then injects its RNA and capsid into the cell; its membranous envelope remains outside the cell.

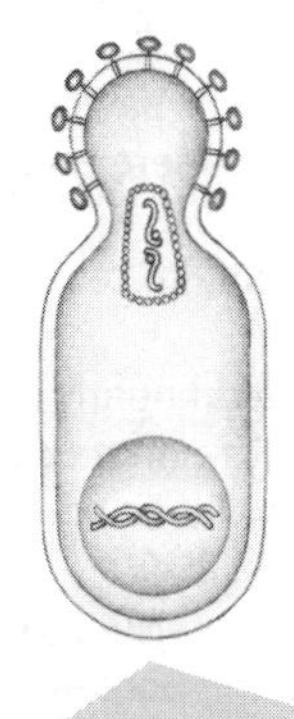

The RNA genome of HIV is transcribed into DNA through the catalytic action of reverse transcriptase.

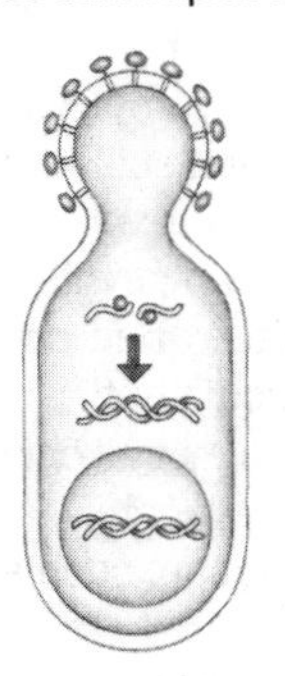

The original RNA genome is destroyed; the new DNA enters the nucleus of the T_4 cell, where it integrates with the T_4 cell's own DNA. The newly integrated DNA is called a provirus.

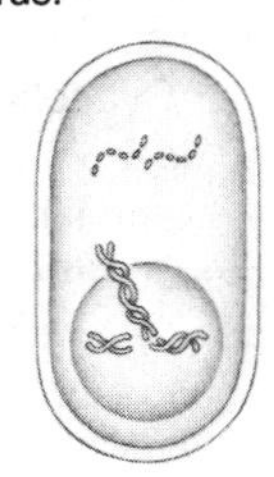

The provirus may remain dormant or, if activated, use the cellular mechanisms of the T_4 cell to manufacute viral RNA. The viral RNA synthesizes proteins, which can be assembled into a new HIV particle.

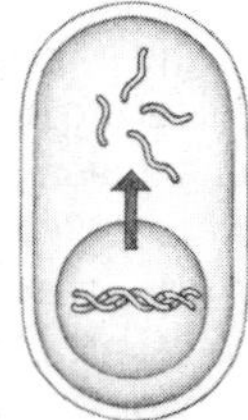

The new HIV particle exits the T_4 cell by budding, often destroying the T_4 cell in the process. The new HIV invades other T_4 cells and the process is repeated.

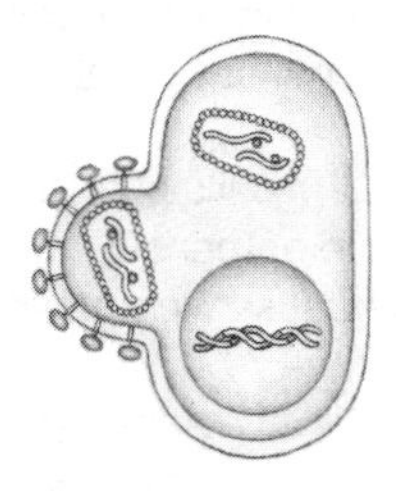

7. Based on a study of molecular systematics (a comparison of amino acids and DNA), scientists have established that prokaryotes split into two divergent lineages —archaebacteria and eubacteria —early in the earth's history

B. Archaebacteria

1. The only surviving archaebacteria are a few genera of prokaryotes that live in harsh environments that resemble earth's early habitats, such as the hot water surrounding the opening of deep-sea vents
2. Archaebacteria differ from eubacteria in the composition of their cell walls and plasma membranes; the plasma membranes of archaebacteria have a lipid composition different from that of any other living organism

C. Eubacteria

1. Eubacteria, or true bacteria, compose the largest group of bacteria, including Cyanobacteria, phototrophic bacteria, pseudomonads, spirochetes, endospore-forming bacteria, enteric bacteria, rickettsiae, chlamydiae, mycoplasmas, actinomycetes, and myxobacteria
2. *Cyanobacteria,* commonly called blue-green algae, are unicellular organisms that are larger than other prokaryotes
   a. These organisms, which typically inhabit fresh water, are photoautotrophs (instead of chloroplasts, they have chlorophyll embedded in thylakoids)
   b. They have no flagella and move by gliding
3. *Phototrophic bacteria* include green sulfur bacteria and purple sulfur bacteria
   a. These photoautotrophs do not release oxygen into the atmosphere as do the Cyanobacteria and other photosynthetic organisms
   b. Most phototrophic species are anaerobic and thrive at the bottom of ponds and lakes and in ocean sediment
4. *Pseudomonads,* the most versatile of all chemoheterotrophs, live in nearly all aquatic and soil habitats; these organisms are able to metabolize organic compounds that no other organism can, such as pesticides and synthetic compounds
5. *Spirochetes* are helical-shaped bacteria that move with a corkscrew-like motion; the bacterium that causes syphilis *(Treponema pallidum)* and the bacterium that causes Lyme disease *(Borrelia burgdorferi)* belong to this group
6. *Endospore-forming bacteria* are rod-shaped organisms capable of creating special thick-walled dehydrated cells (endospores) that enable the organism to withstand the harshest of conditions, even boiling water; *Clostridium botulinum,* which causes botulism, is a type of endospore
7. *Enteric bacteria* are found in the intestinal tracts of animals; some enteric bacteria (such as *Escherichia coli,* which lives in the human intestine) are harmless, whereas others (such as *Salmonella typhimurium,* which causes paratyphoid fever and food poisoning) are harmful
8. *Rickettsiae* are small bacteria capable of living only within the cells of another organism; these parasites, which typically are transmitted to humans by the bites of arthropod vectors (such as ticks and insects), are the causative agents of such diseases as Rocky Mountain spotted fever and typhus
9. Although similar to *Rickettsiae, Chlamydiae* are not transmitted by arthropod vectors; many are transmitted by contaminated genital lesions during sexual intercourse (for example, nongonococcol urethritis is caused this way) or, in the

case of infants, by the vagina during a vaginal delivery (as in chlamydial conjunctivitis)

10. *Mycoplasmas,* the smallest moneran organisms, are unique in that they lack cell walls
11. *Actinomycetes* are organisms that form colonies of branching chains so that they resemble fungi; some actinomycetes species cause tuberculosis and leprosy, whereas others (such as streptomyces) are used to produce antibiotics
12. *Myxobacteria* are gliding bacteria found in the soil; when the soil dries and food becomes scarce, these cells congregate and form a fruiting body —an elaborate colony that is typically brightly colored and measures about 1 millimeter in diameter; the fruiting body then releases spores, which germinate when conditions become more favorable

III. Protista

A. General information

1. Protista were the first eukaryotic organisms to evolve
2. Most, but not all, species are unicellular
3. Protista inhabit water, where they are an important component of plankton and moist terrestrial environments, such as damp soil and leaf litter
4. Some Protista live independently, whereas others are symbionts that inhabit the body fluids and tissues of host organisms
5. Most species are aerobic organisms that use mitochondria for cellular respiration
6. Some species are photoautotrophs that contain chloroplasts; others are heterotrophs
7. At some point in their life cycle, most species use flagella or cilia for movement
8. All protists can reproduce asexually; some also can reproduce sexually through meiotic division
9. A recent modification of the taxonomic classification system has moved multicellular algae and some fungi-like organisms from the Plant kingdom into the Protista kingdom
10. At present, the major Protista groups include protozoa, algal protists, and fungus-like protists (see *Typical Protist Organisms,* page 116, for illustrations of some common varieties)

B. Protozoa

1. Protozoa —one-celled heterotrophs that live primarily by ingesting large organic molecules and other small protists —encompass the phyla Rhizopoda, Actinopoda, Foraminifera, Apicomplexa, Zoomastigina, and Ciliophora
2. *Rhizopoda* (also called *amoeba*) are unicellular organisms that inhabit fresh water, marine environments, and damp soil
   a. These protozoa, which may or may not contain shells, move by means of cellular extensions called ***pseudopodia***
   b. One species, *Entamoeba histolytica,* causes amoebic dysentery in humans
3. *Actinopoda* have slender projections called *axopodia* that help the organism to float and feed

(text continues on page 118)

Typical Protist Organisms

These illustrations show some of the most common features of typical protist organisms.

Dinoflagellata

The outer surface of the dinoflagellata is reinforced by an "armor" of cellulose, which gives the organism its characteristic shape. The dinoflagellata moves (spins) by whipping its two flagella, which protrude from a perpendicular groove along the middle of the cell.

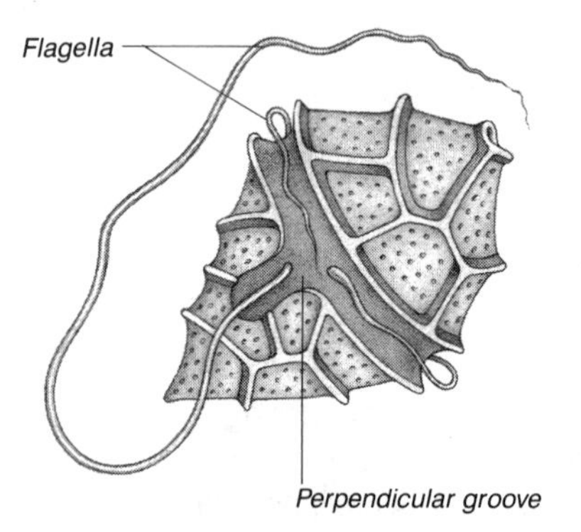

Radiolarian

This organism has a shell made of intricate silicate salts with long, slender extensions called axopodia. The axopodia are composed of microtubules covered by a thin layer of cytoplasm to which smaller protists and other microorganisms become stuck and eventually phagocytized.

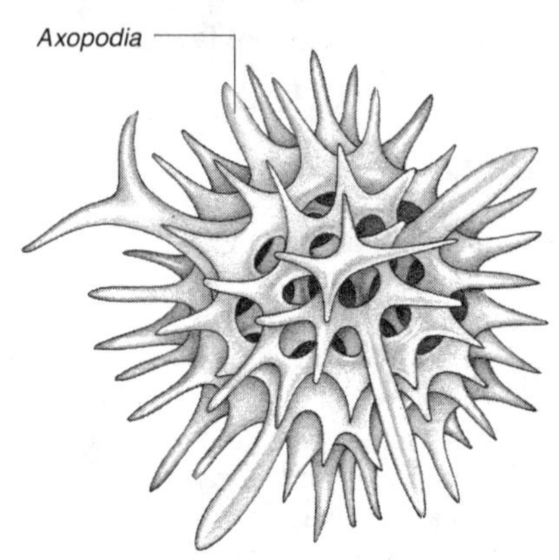

Foraminifera

The shell of this foraminifera is multichambered and contains calcium carbonate. Strands of cytoplasm project from openings in the shell and help the organism to swim, feed, and secrete its shell.

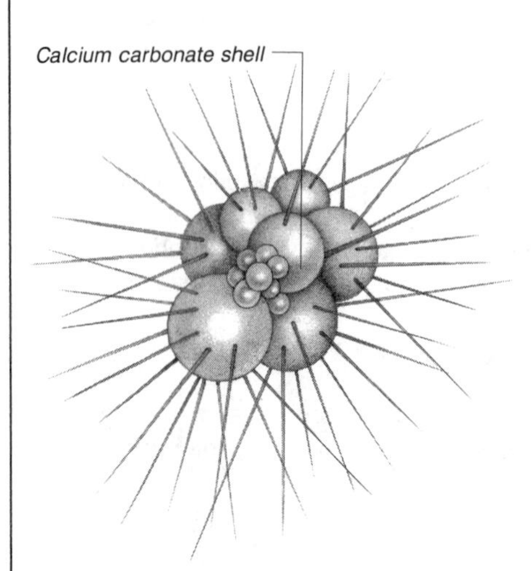

Volvox

A volvox is an amalgamation of individual cells living together in a colony. The center of the ball of cells is hollow. Each cell along the outer wall of the colony is biflagellated, connected to its neighbor by strands of cytoplasm, and embedded in a gelatinous matrix.

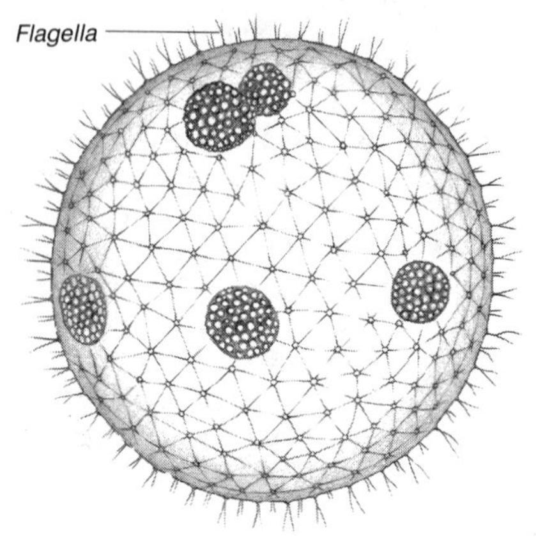

Typical Protist Organisms *(continued)*

Amoeba

A unicellular organism, the amoeba moves and feeds by means of pseudopodia (temporary protrusions that can be extended or retracted). It uses its pseudopodia to engulf food particles (phagocytosis), thereby forming food vacuoles. Water and wastes are excreted from the cell by means of a large contractile vacuole.

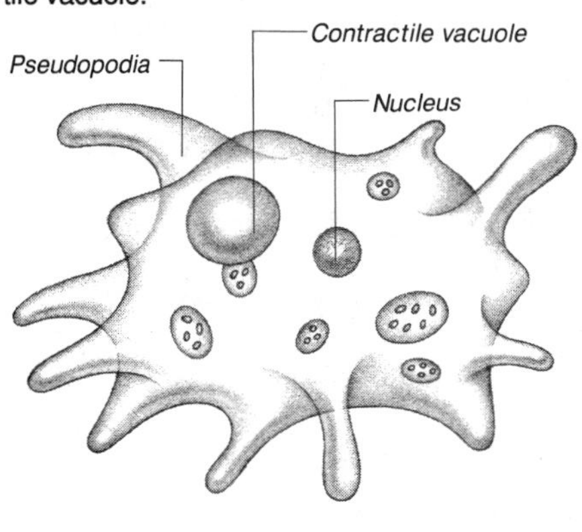

Paramecium

The paramecium uses cilia lining an oral groove (mouth) to sweep in food particles, which are then engulfed by phagocytosis and stored in food vacuoles. Wastes are eliminated from the cell from an anal pore; water is expelled through contractile vacuoles. A macronucleus controls everyday cellular functioning, whereas a micronucleus controls sexual reproduction.

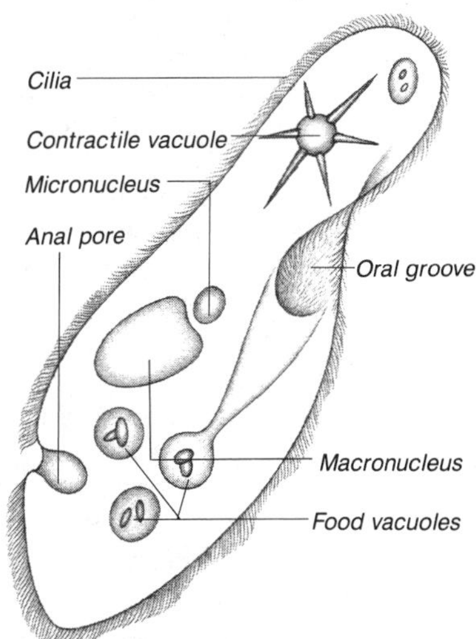

Euglena

This organism has a long flagellum for propulsion and a stigma (a pigment commonly referred to as an eyespot) to guide its movement phototaxically. The euglena lacks a cell wall but has a strong, flexible pellicle (a protein located beneath the plasma membrane) that provides support. Feeding is accomplished in two ways: the euglena contains chloroplasts, which allow it to manufacture food through photosynthesis; it also can gather food through phagocytic action.

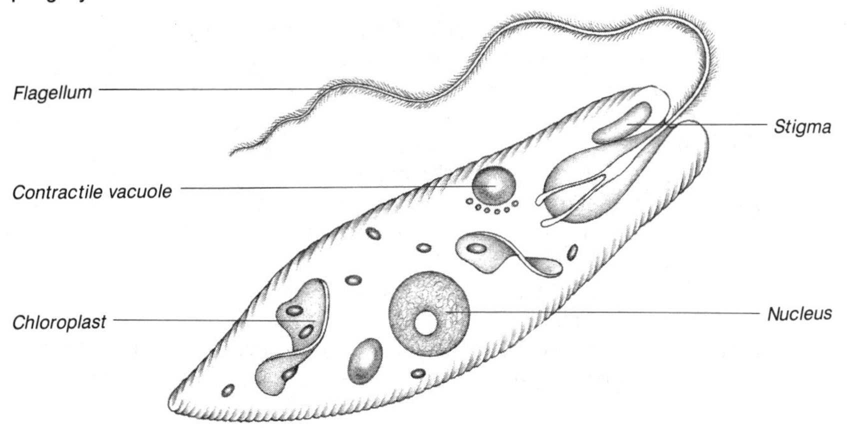

a. These shell-bearing organisms are important constituents of plankton; each species has a different-shaped shell
b. One species (*radiolarians*) is found primarily in marine environments; they have glassy, siliceous shells

4. *Foraminifera* (also called forams) live in marine environments, either in the sand or attached to rocks
   a. These organisms have shells composed of organic material hardened with calcium carbonate
   b. Some species are abundant in plankton
5. *Apicomplexa* (formerly called sporozoans) are animal parasites that colonize their hosts by means of tiny infectious cells called *sporozoites*; the phylum Apicomplexa includes plasmodium, the organism that causes malaria
6. *Zoomastigina* use whip-like flagella to propel themselves
   a. All zoomastigina are heterotrophs; most are unicellular, although some form colonies; they may exist independently or in symbiotic relationships with other organisms
   b. *Trypanosoma,* the organism that causes African sleeping sickness, is a member of this phylum
7. *Ciliophora* use cilia to move and feed; most species (such as *Paramecium*) are unicellular and inhabit fresh water
   a. Unlike other organisms, ciliophora have two types of nuclei: a macronucleus and several micronuclei
   b. They reproduce sexually by means of conjugation

C. Algal protists

1. Most algal protists are photosynthetic organisms; they are classified primarily according to their chloroplast structure, cell wall chemistry, number and position of flagella, and the form of the food stored in their cells
2. *Dinoflagellata* are primarily unicellular with brownish plastids that contain chlorophyll a and c and a mixture of ***carotenoid*** pigments; they are important components of phytoplankton
   a. These organisms, which store food in the form of starch, contain rigid cell walls made of cellulose, which gives them their characteristic shape
   b. Dinoflagellata are propelled by two flagella located in perpendicular grooves within their cell wall
   c. The "blooms" of this organism —indicative of explosive population growths —cause red tides in warm coastal regions
3. *Chrysophyta* (golden algae), a component of freshwater plankton, have yellow and brown carotenoid pigments as well as chlorophyll a and c in their plastids
   a. These flagellated organisms store carbohydrates in the form of laminarin
   b. Their cell walls contain some siliceous material
4. *Bacillariophyta* (diatoms), a component of both marine and freshwater plankton, are nonflagellated organisms that have the same photosynthetic pigments as that of chrysophyta
   a. Their unique, glasslike cell walls are composed of hydrated silica embedded in an organic matrix
   b. Most species are unicellular, although some are filamentous; they store food in the form of an oil, which also provides buoyancy

5. *Euglenophyta* are mostly freshwater, flagellated organisms that contain chlorophyll a and b along with carotenoids
   a. They store food in the form of a polysaccharide called paramylum and have no cell wall
   b. The most well-known organism in this phylum, *euglena,* is capable of photosynthesis; however, if placed in the dark, it also can ingest food particles
6. *Chlorophyta* (green algae) are flagellated organisms that contain chlorophyll a and b along with carotenoids
   a. Unlike other species that inhabit marine environments or damp soil, green algae live mostly in fresh water; some species live symbiotically with fungi in the form of lichens
   b. Green algae store food in the form of starch and have a cell wall made of cellulose
   c. Most chlorophytes are unicellular, although some are multicellular; one multicellular form, *ulva,* is organized into a large body that has the appearance of a leaf
   d. The green algae *volvox* is composed of individual cells that live in a colony
   e. Nearly all chlorophyta reproduce sexually by means of gametes or, in the case of *spirogyra,* by means of conjugation; some green algae exhibit alternation of generations
7. *Phaeophyta* (brown algae), multicellular organisms that primarily inhabit temperate marine waters, compose what is commonly called seaweed; the largest brown algae is known as kelp
   a. These organisms contain chlorophyll a and c along with carotenoids; they store food in the form of laminarin and have a cell wall of cellulose
   b. They possess flagella only during the reproductive stages and reproduce sexually through alternation of generations
8. *Rhodophyta (red algae)* are nonflagellated organisms that contain chlorophyll a and d, as well as carotenoids and an accessory pigment called phycoerythrin
   a. They store food in a form called floridean starch, which resembles glycogen; their cell walls are made of cellulose
   b. Most red algae are marine forms that inhabit tropical waters, but some are freshwater species; all are multicellular and, along with the brown algae, they compose seaweed
   c. Rhodophyta reproduce sexually

D. Fungus-like protists

1. These protists resemble fungi in appearance and life-style but differ from them in terms of cell structure and reproduction
   a. The cell wall of fungus-like protists is made of cellulose; the cell wall of fungi contains chitin
   b. The diploid stage dominates the life cycle of fungus-like protists, whereas the haploid stage dominates the life cycle of fungi
2. *Myxomycota,* or plasmodial slime molds, are heterotrophs; during the feeding stage of the myxomycota's life cycle, the organism forms into an amoeboid mass called a *plasmodium,* which can grow to several centimeters
   a. The plasmodium is a coenocytic mass (a multinucleated continuum of cytoplasm undivided by membranes or walls)
   b. It feeds by means of phagocytosis (flowing over and engulfing solid particles)

c. If the habitat dries up or no food is available, the plasmodium ceases to grow; it then enters the reproductive stage of its life cycle
3. *Acrasiomycota* are cellular slime molds that change shape depending on the food supply
a. When food is available, they exist as single cells
b. When food is scarce, they combine and function as a unit; although the aggregate resembles a slime mold, it is not coenocytic
4. *Oomycota* are commonly called water molds, white rusts, and downy mildews; they resemble fungi in that they have branched filaments, called hyphae, that are coenocytic
a. Most oomycota are saprophytes (heterotrophic organisms that feed on dead plants and animals); some are parasitic
b. Oomycota play a part in decomposition in aquatic environments; they thrive as saprophytes, feeding on dead algae and animals in the water

Study Activities

1. Describe the differences between viruses and bacteria.
2. Compare replication among bacterial, plant, and animal viruses.
3. Draw four protists and identify the structures used for movement.

10

Fungi and Plants

Objectives

After studying this chapter, the reader should be able to:
- List the characteristics of the fungal divisions *Zygomycota, Ascomycota, Basidiomycota,* and *Deuteromycota.*
- Describe three adaptations of plants to terrestrial environments.
- Describe the life-cycle changes that enabled seed-bearing plants to colonize land.
- Explain the major differences between gymnosperms and angiosperms.

I. Fungi

A. General information

1. All organisms within the kingdom Fungi are eukaryotic; although some fungi (such as yeasts) are unicellular, most are multicellular or multinucleate
2. The cells of a multicellular fungus differ from those of a plant or an animal in the number of nuclei and the cell wall structure
   a. Each cell of a fungus is really a nucleated compartment that typically contains more than one nucleus
   b. The partitions (membranes or walls) separating each cell typically are absent, allowing for continuous cytoplasmic movement
   c. The cell walls, which usually appear only during the formation of reproductive structures, are composed of chitin, a nitrogen-containing polysaccharide
3. Fungal cells are characteristically organized into branched filaments called ***hyphae*** that are separated by septa, openings that enable cell-to-cell communication; the branched network of hyphae composing a fungus is called a ***mycelium***
4. Heterotrophic organisms, fungi are primarily saprophytic (feeding on dead organic matter) or parasitic (feeding on a live host)
   a. Saprophytic fungi digest food outside their bodies by secreting hydrolytic enzymes, which decompose complex molecules into simpler compounds for easy absorption
   b. Some parasitic fungi digest food extracellularly, similar to the way saprophytic fungi digest dead organic material; others absorb nutrients directly from their host

5. Some fungi, such as those found in lichens, share a symbiotic relationship with other organisms
6. Fungi may reproduce asexually or sexually
   a. Most fungi are haploid and generate new cells by means of mitosis; when a diploid fungus occurs, it is quickly returned to a haploid state by means of meiosis (thus, fungi also reproduce through alternation of generations)
   b. Sexual reproduction, which involves the union of hyphae of opposite strains (a process called *conjugation*), also can occur
7. Scientists have identified over 100,000 species of fungi, which are categorized according to four major phyla (or divisions): Zygomycota, Ascomycota, Basidiomycota, and Deuteromycota

B. Zygomycota

1. Primarily terrestrial organisms, Zygomycota live in soil or on decaying organic material
2. The hyphae of this type of fungus are coenocytic (without walls); cell walls appear only during the formation of reproductive structures
3. Zygomycota reproduce both asexually and sexually
   a. Asexual reproduction is characterized by the release of spores by special reproductive structures called *sporangia* located at the tips of aerial hyphae
      (1) Typically, one sporangium yields thousands of minute spores that are released upon disintegration of the cell wall
      (2) These spores eventuallly grow into new mycelia
   b. Sexual reproduction, which results from conjugation (the union of two morphologically indistinct cells from hyphae of opposite strains), is characterized by the formation of a zygote
      (1) This process begins when neighboring mycelia of opposite strains form hyphal extensions called *gametangia*
      (2) The gametangia fuse
      (3) The fused gametangia form a thick-walled structure called a *sporangium* containing one or more haploid cells (gametes)
      (4) When conditions are right, the haploid nuclei fuse to produce a zygote
      (5) The zygote undergoes meiosis, producing haploid spores that are released into the environment; each haploid spore develops into a new mycelium
4. The common bread mold, *Rhizopus,* is a member of this division

C. Ascomycota

1. Commonly called *sac fungi,* Ascomycota encompass diverse fungal varieties ranging from unicellular yeasts to complex, multicellular cup fungi
2. Multicellular Ascomycota are characterized by septal partitions between hyphae and by the presence of *asci* (sacs containing sexually produced spores)
   a. The hyphae of multicellular Ascomycota, unlike those of Zygomycota, are partitioned by septa—incomplete cell walls through which cytoplasm circulates and cells communicate

b. During their sexual cycle, Ascomycota produce sacs of spores, or asci (each spore is called an *ascospore;* each separate sac, an *ascus*); groups of asci packed together form structures called *ascocarps* (these compose the cups of a cup fungus)

3. Ascomycota also produce asexual spores (*conidia*) through the mitotic division of cells located at the end of specialized hyphae in the adult haploid fungus
4. The unicellular yeast also undergoes sexual and asexual reproduction
   a. During sexual reproduction, a yeast produces the equivalent of an ascus
   b. During asexual reproduction, the yeast cell divides by budding (the pinching off of a small, new cell from a larger parent cell)
5. Lichens are aggregates of fungi and algae that grow together in symbiosis; the primary fungal component of such aggregates is Ascomycota

D. Basidiomycota

1. This division, which includes some of the largest fungi (mushrooms, shelf fungi, and puffballs), is named for its unique club-shaped reproductive structures — the *basidia*
2. Basidia are haploid hyphae that have fused together to produce a transient diploid organism
   a. The diploid nucleus undergoes meiosis and produces four haploid cells (*basidiospores*)
   b. The basidiospores are dispersed by the wind to produce new adult haploid fungi
   c. Although Basidiomycota reproduce asexually, reproduction by basiodiospores is the distinctive feature of this phylum

E. Deuteromycota

1. Commonly referred to as *imperfect fungi,* Deuteromycota reproduce only asexually; however, if further studies demonstrate that Deuteromycota reproduce sexually, these fungi will be reclassified accordingly
2. Penicillium, from which the antibiotic penicillin is derived, belongs to this division

II. Plants

A. General information

1. All plants are eukaryotic, multicellular organisms with cell walls made of cellulose
2. According to most botanists, plants — the first living organisms to colonize land — evolved from green algae (Chlorophyta)
3. Plants store food in the form of starch; they contain chlorophyll a and b as well as various carotenoid pigments
4. All plants reproduce by mitosis and meiosis, as well as by alternation of generations
5. The evolution of plants from aquatic organisms to primarily terrestrial ones is characterized by the development of structures especially adapted to withstand

excessive water loss, prolonged dryness, temperature fluctuations, and the pull of gravity

a. One adaptation is the development of *vascular tissue* (vascular tissue — xylem and phloem —is made up of cells joined into tubes that transport water and nutrients throughout the plant body)

b. Another adaptation is the development of *seeds* (a seed is a reproductive entity consisting of an embryo and a store of food encased in a protective covering)

c. A third adaptation is the development of *flowers* (the flower is a complex reproductive structure that bears seeds within protective chambers called ovaries)

d. Other adaptations specific to plants that inhabit terrestrial environments include the development of ***stomata*** (openings that allow for the exchange of gases), the *cuticle* (the waxy layer on the surface of leaves and stems that prevents desiccation), and *gametangia* (special organs that protect gametes and embryos from drying out)

6. Plants are commonly categorized into ten major taxonomic divisions (phyla) according to their vascularity and ability to produce seeds (see *Taxonomic Division of Plants)*

B. Nonvascular plants

1. Nonvascular plants lack vascular tissue and must acquire and circulate water through diffusion (movement of a substance from an area of greater concentration to one of lesser concentration), capillary action (a type of cohesive force that pulls water into plants and enables it to rise, much the way water rises in a capillary tube), and cytoplasmic streaming (a phenomenon in which the cytoplasm of a plant cell continuously streams around the space between a vacuole and the plasma membrane)

2. These plants are members of the division Bryophyta, which consists of three classes —mosses, liverworts, and hornworts

a. Mosses have rhizopods (rootlike extensions) that enable them to attach to a substrate, such as a tree or the forest floor, and to absorb and retain water; these plants grow so closely together that they resemble a growing carpet or mat; most mosses grow just a few inches high, but some tropical species may grow to a foot or more; the haploid stage is the dominant generation among these plants

b. Liverworts are less common than mosses; their plant bodies are divided into lobes; the sporangia of liverworts contain special cells called *elaters* (long, slender threads with spirally thickened walls) that help to disperse the spores

c. Hornworts resemble liverworts; however, during the sporophyte generation, the gametophyte grows an elongated, hornlike capsule; another distinctive feature is the presence of one large chloroplast instead of many smaller ones

3. Bryophyta inhabit land but lack many of the terrestrial adaptations of other plants; they were able to colonize land because of two adaptations —a waxy cuticle

Taxonomic Division of Plants

Below are the 10 major divisions (phyla) of plants classified according to their vascularity and ability to produce seeds.

NONVAS-CULAR PLANTS	SEEDLESS VASCULAR PLANTS	SEED-BEARING PLANTS (GYMNOSPERMS)	SEED-BEARING PLANTS (ANGIOSPERMS)
Bryophyta	*Psilophyta*	*Coniferophyta*	*Anthophyta*
	Lycophyta	*Cycadophyta*	
	Sphenophyta	*Gingkophyta*	
	Pterophyta	*Gnetophyta*	

that protects the plant against dessication, and the development of multicellular gametangia that keep the gametes moist

a. They prefer damp and shady habitats

b. They reproduce sexually

(1) The male gametangium of the bryophyta, called an *antheridium,* produces flagellated sperm; the female gametangium, called an *archegonium,* produces one egg

(2) To fertilize the egg, the flagellated sperm must swim from the antheridium to the archegonium; thus the plants must be near water

C. Seedless vascular plants

1. These plants, which include the divisions Psilophyta, Lycophyta, Sphenophyta, and Pterophyta, contain vascular tissue through which water and nutrients are transported

2. Psilophyta are the most primitive variety of vascular plants

a. They lack true roots and leaves and resemble ferns

b. *Psilotum,* commonly called the whiskfern, is the only extant (surviving) member of this division

3. Lycophyta include club mosses, which grow on forest floors, and many tropical species of ***epiphytes*** (plants that nourish themselves but grow on the surface of another plant for support); they reproduce sexually, although the methods can vary among different species

a. In all species, the gametophyte stage is inconspicuous; the gametophyte is a dependent organism that lives underground and obtains nourishment from symbiotic fungi

b. In some species, a single gametophyte produces both male and female gametes; these bisexual gametophytes arise from a single type of spore; the sporophyte that produces this spore is considered homosporous

c. Other species have two gametophytes —a male that produces only male gametes and a female that produces only female gametes; each gametophyte arises from a different type of spore (the male gametophyte from a microspore, the female gametophyte from a megaspore); the sporophyte that produces these two different types of spores is considered heterosporous

d. Club mosses have specialized leaves (sporophylls) for reproduction; in some species, these leaves are bunched at the tips of branches, forming a club-shaped structure called a *strobilus;* club mosses also have leaves with strands of vascular tissue

4. Sphenophyta (commonly called horsetails) contain minute gametophytes that are photosynthetic and free-living (not dependent on the sporophyte for food); only about 15 species of Sphenophyta still exist
   a. The sporophyte has an underground rhizome from which aerial stems arise; these stems have whorls of small leaves emerging from jointed sections; in some species, the stems are evergreen and perennial
   b. The epidermis is embedded with silica, which gives the plant an abrasive texture
   c. Most horsetails grow 1 to 3 feet; some tropical plants grow much taller
5. Pterophyta —the most prevalent of all seedless plants with more than 12,000 species, including ferns —possess leaves that are much larger than those of other seedless plants
   a. These large leaves, called *fronds,* contain a branched system of veins
   b. Some fronds are specialized into sporophylls (leaves that bear sporangia)
   c. A cluster of sporangia is called a ***sorus***

D. Seed-bearing plants

1. The most prevalent type of terrestrial plants, seed-bearing plants were most successful at colonizing land as a result of three evolutionary changes related to life-cycle events
   a. Gametophytes evolved from separate, free-living organisms into minute organisms capable of living within the moist reproductive tissue of sporophytes, where they were protected from dessication
   b. Pollination enabled the delivery of sperm to eggs by air rather than by water
   c. The seed —a reproductive entity consisting of an embryo and stored food encased in a protective coating —provided a nutritious environment in which an embryonic plant can remain dormant for long periods until unfavorable conditions (such as drought or cold) became more favorable for germination and dispersement over land
2. Seed-bearing plants are categorized primarily into two groups —***gymnosperms*** and angiosperms —based on the appearance and placement of reproductive structures
   a. The reproductive structures of gymnosperms (*cones*) are found on mature sporophytes (trees); the cones contain sporangia and the developing gametophytes
      (1) Many gymnosperm species possess two types of cones: *pollen cones,* which contain male gametophytes that develop from microspores,

and *ovulate cones,* which contain female gametophytes that develop from megaspores

(2) Pollination occurs when the male gametes from the pollen cones are carried by the wind to the ovulate cones

(3) After fertilization, the seed develops on the ovulate cone and eventually disperses to grow into an adult sporophyte

b. The reproductive structures of angiosperms are found within the flowers of sporophytes

(1) The sporophyte produces male gametophytes from microspores and female gametophytes from megaspores; the gametophytes remain within the structure of the flower

(2) Pollination occurs when the male gametes, which are released from the anther, land on the stigma

(3) After fertilization, the seed remains within the flower's ovary, which may ripen into a fruit

3. *Gymnosperms*—the earliest seed plants to appear in the fossil record—encompass several divisions

a. Coniferophyta (conifers), the largest division, include about 550 known species (such as pines, firs, spruce, cedars, cypresses, and redwoods)

(1) Nearly all conifers are evergreens, which means that they retain their leaves throughout the year

(2) The needle-shaped leaves of pines and firs are especially adapted to dry conditions; a thick cuticle covers the leaf, and the stomata are located in pits, which reduces water loss

b. Cycadophyta encompass about 100 known species, including the cycads; these plants, which grow to 30 feet, resemble palms but do not flower; they bear naked seeds on the scales of cones; and they are dioecious (pollen and seeds are produced on different trees)

c. Ginkgophyta include only one extant species, the gingko tree; also known as the maidenhair tree, the gingko has fernlike leaves that turn gold and shed in the fall; this dioecious plant may grow to 100 feet tall

d. Gnetophyta include about 70 species, among which are the ephedra, a small shrub that grows in the North American deserts, and the welwitschia,a plant with straplike leaves found in the deserts of southeastern Africa

4. *Angiosperms*—members of the Anthophyta division—include about 235,000 different species of flowering plants; the two main classes within this division are the monocotyledons (monocots) and dicotyledons (dicots)

a. *Monocots* contain one cotyledon (seed leaf); they typically possess leaves with parallel veins, vascular bundles arranged in complex patterns, a fibrous root system, and flowers whose main parts occur in multiples of three; examples include lilies, orchids, grasses, and grain crops

b. *Dicots* contain two cotyledons; they typically have leaves with netlike veins, vascular bundles arranged in a ring, and taproots and floral parts that usually occur in multiples of four or five; examples include roses, peas, sunflowers, oaks, and maples

Study Activities

1. Outline the methods of reproduction in the four fungal divisions.
2. Draw a fungus cell and a plant cell, labeling the major structures and organelles.
3. Compare water transport in vascular and nonvascular plants.
4. Describe reproduction in nonvascular, seedless vascular, and seed-bearing plants.
5. List the major characteristics of monocots and dicots, providing three examples of each class.

11

Animals

Objectives

After studying this chapter, the reader should be able to:
- Describe five characteristics common to all animals.
- Outline the key points of divergence in the evolutionary history of animals.
- Characterize seven of the major phyla of the animal kingdom.
- Discuss the phases of vertebrate evolution.
- Compare the circulatory, respiratory, and reproductive features of the seven classes of vertebrates.

I. Animal Kingdom

A. General information

1. All animals are multicellular, eukaryotic organisms with cell membranes (not cell walls); they inhabit all types of aquatic and terrestrial environments
2. Heterotrophic organisms, animals ingest food for nutrition; their carbohydrate reserves are stored as glycogen
3. All animals except for the simplest variety (sponges) possess a neuromuscular system containing a network of nerves and muscles
4. Animals typically reproduce by sexual means; the diploid stage usually is the predominant stage during an animal's life cycle
5. The kingdom Animalia is divided into two subkingdoms: ***Parazoa*** and ***Eumetazoa***
    a. Parazoa comprises only one phylum of animals —Porifera, commonly known as sponges; although most biologists believe that all animals evolved from one group of protists, the unique development and anatomic simplicity of the Parazoa set them apart from other animals
    b. Eumetazoa comprises all of the remaining phyla within the animal kingdom
6. Animals within the subkingdom Eumetazoa are further divided into a system of branches and subbranches, each of which represents a major point at which animals diverged during the course of evolution
7. The evolutionary changes marking key divergences include a change in body symmetry, the development of a true body cavity, changes in embryonic development, and the development of a notochord (see *Branches of the Animal Kingdom*, page 131, for a depiction of the evolutionary progression of one branch to another)

B. Branches based on changes in body symmetry

1. The subkingdom Eumetazoa is divided into two major branches —***Radiata*** and ***Bilateria***—based on body symmetry
2. The branch Radiata includes all animals with *radial symmetry* (possessing a top and a bottom, but no definitive front or back or left or right side)
3. The branch Bilateria includes all animals that currently have *bilateral symmetry* (a top and a bottom that includes a head [anterior] end and a tail [posterior] end) as well as all animals that are believed to have evolved from an ancestor with bilateral symmetry
    a. All animals classified as Bilateria are characterized by an evolutionary trend toward cephalization (a concentration of sensory equipment on the anterior end) and the presence of three embryonic germ tissues (ectoderm, mesoderm, and endoderm), which give rise to all the various tissues and organs of the adult organism
    b. Some animals classified as Bilateria, such as sea stars (phylum Echinodermata), are radially symmetrical as adults; they are placed within the branch Bilateria because their embryonic development and internal anatomy suggest that they evolved from an ancestral bilateral form

C. Branches based on development of a true body cavity

1. The branch Bilateria is divided into three subbranches —***Coelomates***, ***Pseudocoelomates***, and ***Acoelomates***—based on the presence or absence of a *coelom* (a body cavity between the gut and the outer body wall that is completely lined with mesoderm)
2. Coelomates have a coelom, or a "true" body cavity completely lined with mesoderm
3. Pseudocoelomates have a coelom that is only partially lined with mesoderm
4. Acoelomates have no coelom

D. Branches based on changes in embryonic development

1. Animals within the branch Coelomates are further divided into three subbranches —***Deuterostomes***, ***Protostomes***, and Lophophorates —based on differences in their embryonic development
2. Deuterostomes and Protostomes —the major evolutionary branches of Coelomates —differ from one another in three ways
    a. First, embryonic cleavage may be spiral or radial
        (1) In Deuterostomes, the dividing zygote undergoes *radial cleavage,* in which the planes of cell division are parallel or perpendicular to the vertical axis of the embryo (see the illustration below)

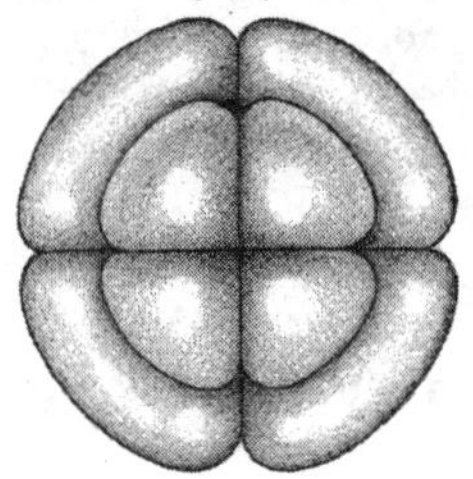

Branches of the Animal Kingdom

Based on comparative anatomy and embryological data, biologists have identified the following major branches within the animal kingdom. The shaded areas indicate major evolutionary trends that led to the development of these branches.

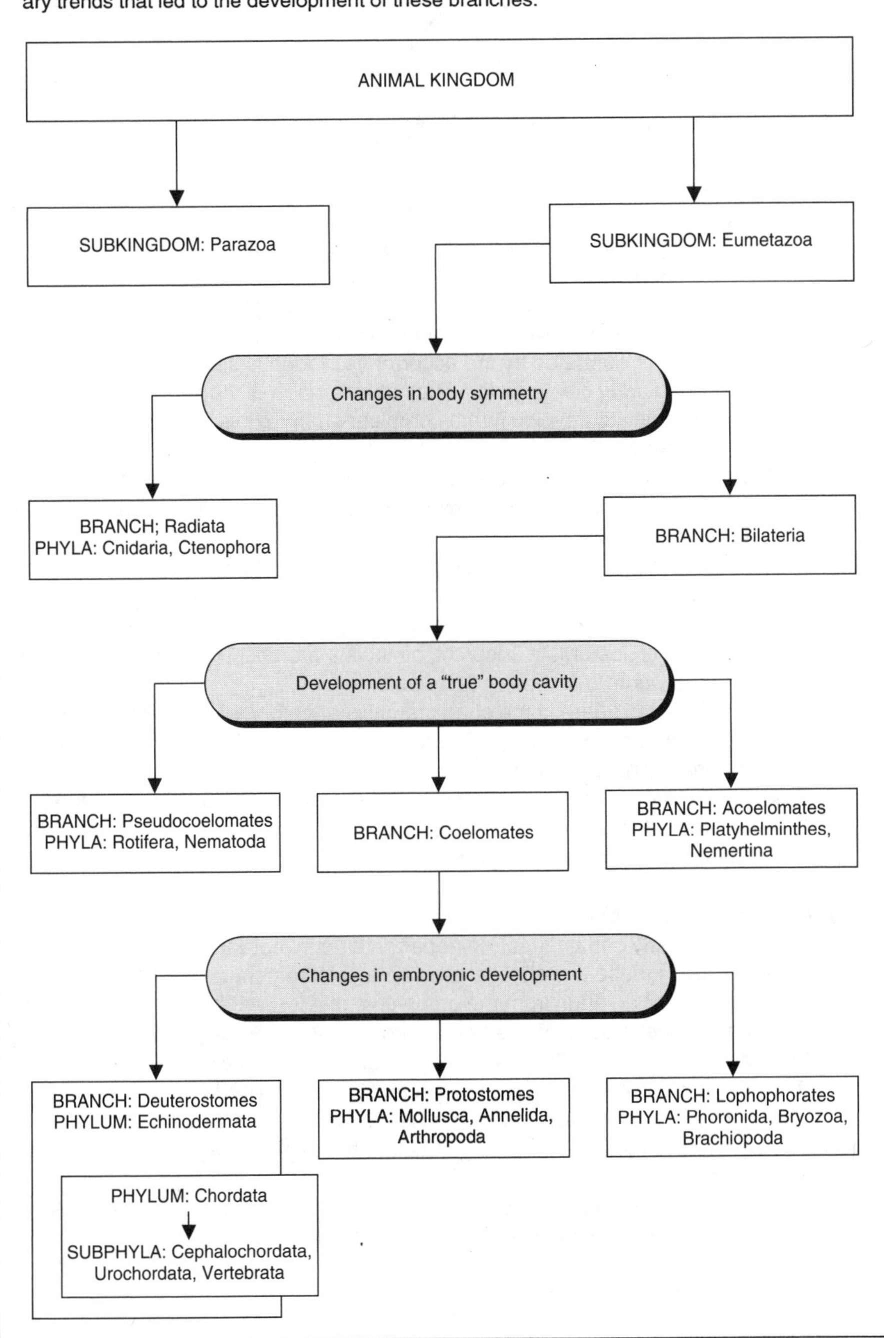

(2) In Protostomes, the dividing zygote undergoes *spiral cleavage,* in which the planes of cell division are diagonal to the vertical axis of the embryo (see the illustration below)

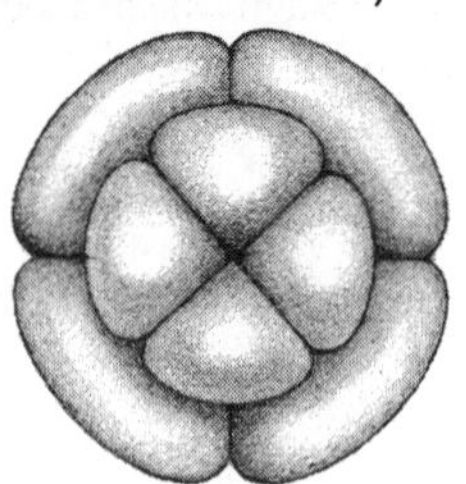

b. Second, embryonic cleavage may be determinate or indeterminate
(1) During early cleavage in a Deuterostome, each of the zygote's cells can develop into a complete embryo if separated from the others (a phenomenon called *indeterminate cleavage*); in humans, this phenomenon is illustrated by the occurrence of identical twins
(2) During early cleavage in a Protostome, each of the zygote's cells loses its ability to develop into a complete embryo; this is called *determinate cleavage*

c. Third, the *blastopore*—an opening formed during the gastrula stage that connects the embryo with its external environment—may develop into a mouth or an anus
(1) In Deuterostomes, the blastopore develops into an anus
(2) In Protostomes, the blastopore develops into a mouth

3. Lophophorates are a taxonomic puzzle; although these animals clearly descended from a Coelomate ancestor, biologists are unsure of their evolutionary affinity to the Deuterostomes and Protostomes
   a. Lophophorates share some characteristics of embryonic development with the Deuterostomes and the Protostomes, including a true body cavity, determinate cleavage, and bilateral symmetry
   b. Unlike the other Coelomates, Lophophorates have a distinctive structure called the *lophophore*—a horseshoe-shaped feeding organ that contains ciliated tentacles and surrounds the mouth

E. Development of a notochord

1. Scientists are fairly certain that Echinodermata and Chordata stem from a common ancestor; these two phyla share some common characteristics, including the formation of an anus from the embryonic blastopore, intermediate cleavage, and the development of a coelom from a mesodermal pouch
2. At some point in the evolutionary process, the Echinodermata deviated greatly from the ancestral type by producing a radial adult from a bilateral larva and by failing to develop a ***notochord***, which is characteristic of all ***Chordates***
3. The notochord is the first supporting structure to develop in a Chordate embryo; it forms above the primitive gut
   a. In turnicates (members of the subphylum Urochordata), the notochord is present only in the tail and only during the larval development

b. In lancelets (members of the subphylum Cephalochordata), the notochord extends almost the entire body length

c. In fishes, amphibians, reptiles, and mammals (members of the subphylum Vertebrata), the notochord is surrounded by tissue and eventually develops into a vertebral column

II. Animal Phyla

A. General information

1. Biologists have identified about 35 different animal phyla, only one of which is classified in the subkingdom Parazoa — Porifera
2. All other animal phyla are classified as belonging to the subkingdom Eumetazoa

B. Porifera

1. Commonly called sponges, Porifera are sessile organisms whose bodies, which resemble sacs filled with holes or pores, remain sedentary and attached by a base to a submerged object
2. Of the approximately 5,000 species of sponges, only about 100 live in fresh water
3. Although sponges lack neuromuscular systems, their individual cells can sense and react to the environment
4. Sponges are filter feeders, collecting food particles from water streaming through them
   a. Water is drawn through the sponge's pores into a central cavity called a *spongocoel* (see *Anatomy of a Sponge,* page 134)
   b. Water flows out of the sponge through a large opening called an *osculum*
5. The wall of a sponge is composed of two layers separated by a gelatinous matrix called the *mesophyll*
   a. The outer layer is composed of epidermal cells
   b. The inner layer contains special cells called *choanocytes,* which line the spongocoel; choanocytes contain a beating flagellum that pulls water through the sponge and a collar of fingerlike projections that trap food particles
   c. Amoebocytes — special cells that wander through the mesophyll — take up food from the choanocytes, then digest the food and carry it to the epidermal cells; they also secrete tough skeletal fibers called *spicules*

C. Cnidaria

1. Formerly called Coelenterata, this phylum includes more than 10,000 species consisting of hydras, jellyfish, sea anemones, and corals
2. The cnidarian's saclike body contains a central digestive compartment called a *gastrovascular cavity* and a single opening that serves as both a mouth and an anus
3. These animals have rudimentary muscles and nerves; cells of the epidermis and gastrodermis contain microfilaments that can contract and produce slow movement, which is coordinated by a nerve net (a nerve net is the simplest type of nervous system; it has no central control, such as a brain or an enlarged ganglion)

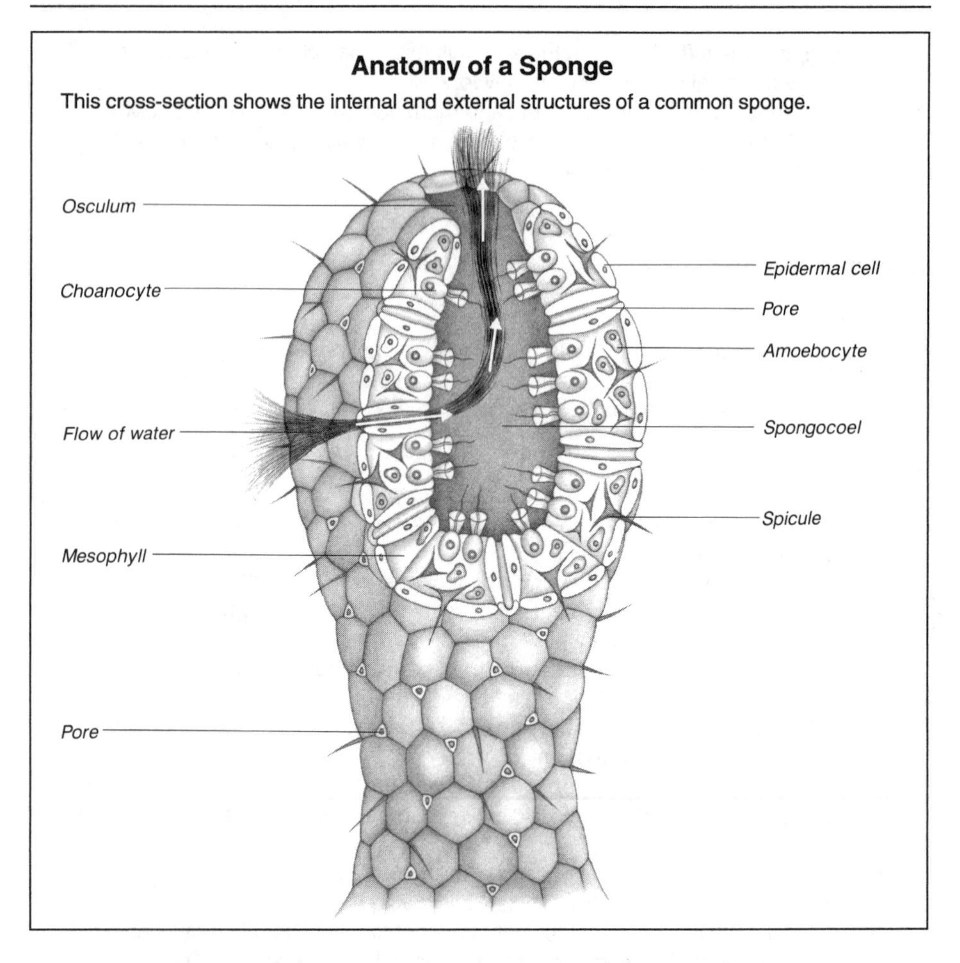

4. Cnidarians are carnivores that use the tentacles surrounding their single opening to capture food and push it into the gastrovascular cavity; the undigested remains are egested through the same opening
5. The tentacles contain *cnidocytes,* which are special epidermal cells that defend the animal and help capture prey; each cnidocyte contains a stinging capsule, or *nematocyst*
6. Biologists have identified two forms of cnidarians —polyps and medusae
   a. The *polyp* cnidarian is a sessile form that attaches to a substratum by means of a body stalk and waves its tentacles waiting for prey; the hydra is an example of a polyp
   b. The *medusa* cnidarian is a free-moving form that propels itself by contracting its muscle cells; the jellyfish is an example of a medusa

D. Ctenophora

1. This phylum, commonly called comb jellies, consists of about 100 species of marine organisms
2. Ctenophores contain eight rows of cilia that are used for locomotion

3. They have a sensory organ made up of calcareous particles; the settling of these particles initiates a nerve impulse that directs the cilia to coordinate locomotion
4. Ctenophores have tentacles but no cnidocytes and therefore cannot paralyze their prey

E. Platyhelminthes

1. Commonly called flatworms, this phylum includes more than 15,000 species
2. Flatworms have a gastrovascular cavity with only one opening
3. Platyhelminthes are divided into three classes: Turbellaria, Trematoda, and Cestoda
   a. Animals belonging to the class Turbellaria are mostly nonparasitic and marine, although species collectively referred to as planarians live in fresh water
      (1) Planarians are carnivores that move by using cilia on the ventral epidermis to glide along a film of mucus; they have a pair of eyespots that detect light and lateral flaps called *auricles* that function in smell
      (2) They have a rudimentary brain and can learn to respond to stimuli
      (3) Planarians are hermaphrodites that reproduce both sexually and asexually through regeneration; in regeneration, the parent constricts its middle, cuts itself in half, and then grows back the missing ends
   b. Animals belonging to the class Trematoda are parasites commonly called flukes
      (1) Flukes use suckers to attach to the host
      (2) The blood fluke *Schistosoma* infects humans, causing schistosomiasis
   c. Animals belonging to the class Cestoda are parasites; an example is the tapeworm

F. Nemertina

1. Consisting of about 900 mostly marine species, Nemertina are commonly called proboscis worms
2. They have a long, retractable hollow tube called a *proboscis* that is used to sense objects, defend against predators, and capture prey
3. These animals have a complete digestive tract, including separate openings for a mouth and an anus, and a simple circulatory system containing red blood cells but no heart

G. Rotifera

1. Rotifera comprise about 1,800 species inhabiting fresh water and damp soil
2. These miniscule (no more than 2 mm long) animals have a complete digestive tract
3. They possess a distinctive wheel-shaped organ consisting of a crown of cilia that draws water into the mouth
4. Some species consist of only females that reproduce by parthenogenesis —that is, producing offspring from unfertilized eggs

H. Nematoda

1. Commonly called roundworms, the 80,000 species of this phylum live in water, moist plant tissues, wet soil, and the body fluids of other animals; those that live in soil play an key role in decomposition and nutrient cycling

2. Roundworms possess a complete digestive tract and range in size from 1 mm to over 1 m in length
3. Most species of Nematoda have separate sexes and reproduce sexually
4. About 50 different species parasitize humans, including *Trichinella spiralis,* which causes trichinosis in humans

I. Phoronida

1. These marine worms range from 1 to 50 mm in length and live in the sand
2. Phoronids have tubes made of chitin and extend their lophophores from the openings of the tubes
3. Hermaphrodites, the Phoronids are also capable of asexual reproduction

J. Bryozoa

1. These tiny animals live in colonies, mostly in marine habitats
2. In most species, the colony is covered by a hard exoskeleton containing tiny pores through which the animals extend their lophophores
3. Bryozoans are integral to reef building
4. Hermaphrodites, Bryozoans are also capable of asexual reproduction by budding

K. Brachiopoda

1. Also called lampshells, Brachiopods are marine organisms that resemble clams
2. These solitary animals attach to a substratum by means of a stalk and open their shells slightly to extend their lophophores
3. The sexes are separate; the egg and sperm are discharged into the sea, and the larvae are ciliated and free-swimming

L. Mollusca

1. This phylum includes more than 100,000 known species of snails, slugs, oysters, clams, squid, and octopuses
2. Although most species are marine, some inhabit fresh water and others (such as some species of snails) live on land
3. Mollusks are mostly soft-bodied animals protected by a shell (some species of squid and octopus have diminished shells; some lost their shells during their evolution)
4. All mollusks have a muscular foot (used for movement), a visceral mass (containing most of the internal organs), and a mantle (a heavy fold of tissue that covers the visceral mass and that secretes the shell in those species with shells)
5. Most mollusks feed by means of a rough, muscular tongue called a *radula*
6. Except for garden snails, all species of mollusks have separate sexes
7. The only mollusks with a closed circulatory system belong to the class Cephalopoda (squid and octopus); cephalopods also have a well-developed nervous system with a complex brain

M. Annelida

1. The approximately 10,000 species of annelids live in seawater, fresh water, and damp soil
2. Annelids have segmented bodies and range from 1 mm to 3 m in length
3. Their complete digestive system includes such specialized organs as a pharynx, an esophagus, a crop, a gizzard, and intestines

4. Annelids have a closed circulatory system
5. The most well-known annelid, the earthworm, belongs to the class Oligochaeta
   a. The earthworm's nervous system consists of a pair of cerebral ganglia connected to ganglia present in each body segment
   b. Earthworms are hermaphrodites, but they cross-fertilize

N. Arthropoda

1. Arthropods, which include nearly 1 million species, have jointed appendages modified for walking, feeding, sensory reception, copulation, and defense
2. The arthropod's body is covered by an external skeleton (***exoskeleton***) composed of protein and ***chitin***
   a. The exoskeleton protects the animal and serves as a point of attachment for muscles
   b. To grow, the arthropod must shed its old exoskeleton and secrete a larger one—a process known as ***molting***
3. Arthropods have well-developed sensory organs, including eyes, olfactory receptors, and antennae for touch and smell
4. Cephalization is extensive—that is, the sensory and feeding organs are concentrated at the anterior end of the animal
5. Arthropods have an open circulatory system; a fluid called *hemolymph* leaves the heart and passes into open spaces (sinuses) surrounding the tissues and organs; the sinuses (called *hemocoels*) are not part of the coelom
6. Respiration in aquatic species is accomplished by gills; in terrestrial species, it is accomplished by branched air ducts that make up the tracheal system
7. Most arthropod species are found within five classes: Arachnida, Crustacea, Chilopoda, Diplopoda, and Insecta
   a. Arachnida include scorpions, spiders, ticks, and mites; these animals have six pairs of appendages: one pair of chelicerae (which aid in feeding), one pair of pedipalps (which also aid in feeding), and four pairs of legs
   b. Crustacea include lobsters, crayfish, crabs, copepods, barnacles, and shrimp; these animals, which may have up to 20 pairs of appendages, are the only arthropods with two pairs of antennae
   c. Chilopoda include centipedes, which are terrestrial carnivores; each segment of their trunk region has one pair of legs
   d. Diplopoda include millipedes, which feed on decaying leaves and other plant matter; each segment of their trunk region contains two pairs of legs
   e. Insecta, the largest class of arthropods, include about 26 different orders
      (1) Insects have three pairs of legs and one pair of antenna; the insect body is divided into the head, thorax, and abdomen
      (2) Most insects have one or two pairs of wings attached to the dorsal side of the thorax; these wings are extensions of the endoskeleton and are not true appendages
      (3) Most insects undergo metamorphosis (a transition from larval to adult form)
      (4) Insects usually reproduce sexually between separate sexes; fertilization occurs internally

O. Echinodermata

1. This phylum, which comprises sea stars, sea urchins, sand dollars, brittle stars, sea lilies, and sea cucumbers, is made up of mostly sessile organisms that display radial symmetry as adults
2. Echinoderms have a thin skin covering an ***endoskeleton*** of hard calcareous plates
3. All echinoderms possess a distinctive organ—a water vascular system, or a network of hydraulic canals ending in extensions called *tube feet;* this system functions in locomotion, feeding, and gas exchange
4. Echinoderms have two sexes and reproduce sexually; fertilization occurs externally
5. Sea stars and other echinoderms have strong powers of regeneration

P. Chordata

1. All chordates are distinguished by the presence of four anatomic structures during embryonic development: a notochord (a flexible rod that runs longitudinally between the gut and nerve cord); a hollow, dorsal nerve cord; pharyngeal slits; and a postanal tail (a tail that extends beyond the anus)
2. In many adult chordates, some of these structures have been modified, diminished, or eliminated
3. Chordata are divided into three subphyla: Cephalochordata, Urochordata, and Vertebrata
   a. Cephalochordata include tiny, marine animals known as *lancelets;* these animals possess a notochord, a dorsal hollow nerve cord, pharyngeal slits, and a postanal tail in both the adult and embryonic forms
   b. Urochordata include animals collectively referred to as *tunicates*; these animals have both a larval and an adult stage
      (1) During the larval stage, they are free swimmers; they retain all the chordate embryonic features
      (2) During adulthood, they become sessile and lose all the chordate embryonic features except for pharyngeal slits
   c. Vertebrata vary in the extent to which the embryonic chordate features are retained in the adult stages
      (1) All vertebrates possess a feature not seen in other subphyla—a vertebra (backbone), which is a bony covering enclosing the nerve cord
      (2) Most vertebrates also have an appendicular skeleton and a closed circulatory system with a two- to four-chambered heart

III. Classes of Vertebrates

A. General information

1. Vertebrates encompass seven classes of animals, including Agnatha (jawless vertebrates), Chondrichthyes (cartilaginous fishes), Osteichthyes (bony fishes), Amphibia (amphibians), Reptilia (reptiles), Aves (birds), and Mammalia (mammals)
2. The first vertebrates to appear in the fossil record, jawless fishes gave rise to the classes Agnatha, Placodermi (now extinct), and Chondrichthyes, as well as primitive ancestral bony fishes

3. The ancestral bony fishes gave rise to the classes Osteichthyes and Amphibia and a group of ancestral reptiles that gave rise to various reptilian orders
4. From the ancestral stem reptiles evolved the classes Reptilia and Mammalia and the dinosaurs
5. Before their extinction, dinosaurs gave rise to the class Aves

B. Agnatha

1. The only living species of this class are the lampreys and hagfishes
2. These jawless animals have a cartilaginous skeleton and a notochord that persists throughout life; they lack the paired appendages found in other classes
3. The eel-shaped adult sea lamprey feeds by attaching itself to the flank of a live fish and using its rasping tongue to penetrate the skin of its prey and feed on its blood; lampreys live as filter-feeding larvae in fresh water, then migrate to sea or lakes where they mature into adults
4. Hagfishes are primarily scavengers; they lack a larval stage and live in salt water

C. Chondrichthyes

1. This class includes sharks and rays, which have a cartilaginous skeleton, paired fins, and biting jaws with sharp teeth derived from jagged skin scales
2. Along each side of the shark is a row of microscopic organs (a lateral line system) sensitive to changes in surrounding water pressure; these organs enable the shark, which is carniverous, to detect minor vibrations and seek out prey
3. Sharks reproduce through internal fertilization; most species are oviparous or ovoviviparous, whereas few are viviparous
4. The shark's reproductive tract, excretory system, and digestive tract empty into a common chamber (the cloaca), which vents to the outside by a single opening
5. Sharks must swim continuously to move water into the mouth and out through the gills; this is necessary for respiration and buoyancy

D. Osteichthyes

1. This class includes the most species of all vertebrates, including all species of fresh water and marine fishes (such as trout, bass, perch, and flounder)
2. Osteichthyes have bony skeletons and a slimy skin that reduces friction as they swim
3. These animals have a protective flap, called an *operculum,* located over their gills
    a. The operculum and some muscles contained within the gill chambers enable bony fishes to actively pump water into their mouth, through the pharynx, and out between the gills
    b. This adaptation enables them to breathe while stationary
4. A special air sac called a *swim bladder* allows bony fishes to adjust their buoyancy so that they can remain motionless in the water
5. Most species are oviparous, reproducing by external fertilization after the female sheds large numbers of eggs; some species reproduce through internal fertilization

E. Amphibia

1. The first vertebrates to begin the transition from water to land, amphibians include such animals as salamanders and frogs

2. Most amphibians undergo a larval stage that is spent in the water and an adult stage that is adapted to living on land
   a. In the frog, the tadpole (larval stage) is an aquatic herbivore with internal and external gills and a lateral line system
   b. The adult frog is a carnivore with developed legs and air-breathing lungs
3. Despite the transition to land, most adult amphibians must remain close to water
   a. Amphibians have inefficient lungs and must keep their skin moist to allow gases to diffuse in and out
   b. Because their eggs have no shell and dehydrate quickly in air, amphibians must reproduce in a moist environment; fertilization occurs externally

F. Reptilia

1. Members of this class (which includes lizards, snakes, turtles, and crocodiles) were the first vertebrates to adapt completely to life on land
2. Reptiles have scales containing the protein keratin
   a. The scales prevent dehydration and waterproof the skin
   b. Because scales also prevent any exchange of gases, reptiles must rely completely on their lungs for respiration
3. Most species lay eggs filled with amniotic fluid on land; the egg's shell prevents desiccation
4. Fertilization occurs internally, before the shell is secreted
5. Reptiles are ectothermic (without a constant body temperature) but can regulate body temperature through behavioral adaptations, such as basking in the sun and seeking shade

G. Aves

1. Birds are characterized by their many physical adaptations that enable them to fly, including honey-combed bones (which are strong but light), wings, minimal organs (females have only one ovary and all birds lack feet—features that reduce overall weight), excellent eyesight, a constant (endothermic) body temperature, feathers and a layer of fat (to provide insulation), a four-chambered heart that separates oxygenated from deoxygenated blood, and efficient lungs
2. Birds' brains are larger than those of reptiles and amphibians
3. Feathers are made from keratin and are thought to have evolved from reptilian scales
4. Birds do not chew food but pass it directly to their gizzard—a digestive organ near the stomach that grinds the food
5. Birds lay eggs, which are fertilized internally

H. Mammalia

1. All mammals have hair, mammary glands to nurse their offspring, a constant (endothermic) body temperature, a diaphragm (a muscle between the thorax and abdomen to aid in ventilating the lungs), and a four-chambered heart that separates oxygenated from deoxygenated blood; they also undergo internal fertilization
2. Mammals are divided into three major groups: the monotremes (egg-laying mammals), marsupials (mammals with pouches), and placental mammals
   a. *Monotremes* comprise the platypus and the spiny anteater—the only living representatives of this group

b. *Marsupials* include opossums, kangaroos, and koalas; these animals complete their embryonic development in a marsupium, or maternal pouch, located outside the female reproductive tract

c. *Placental mammals*—the group that includes humans—complete their embryonic development within the female reproductive tract; the embryo is joined to the mother by the placenta

Study Activities

1. Outline the major differences between deuterostomes and protostomes.
2. Compare the digestive features of cnidarians, platyhelminthes, nematodes, and annelids.
3. Describe the skeletal features of arthropods, echinoderms, and vertebrates.
4. Compare the external anatomy of the five classes of arthropods.
5. Identify four features common to all mammals.
6. Describe the three types of embryonic development in mammals.

12

Organization and Function in Vertebrates

Objectives

After studying this chapter, the reader should be able to:

- Describe the organization and function of the circulatory system in vertebrates.
- Outline how blood clotting is achieved.
- List the differences between aquatic and terrestrial gas-exchange organs in vertebrates.
- Describe the organization and function of the digestive system in vertebrates.
- Identify the basic structures that constitute the nervous system.
- List the sequence of events involved in nerve impulse transmission.
- Describe the organization and function of the human brain.
- Name the structures and processes involved in vision, hearing, and balance.
- Describe the structure and organization of the musculoskeletal system in vertebrates.
- Outline the molecular basis of muscle contraction.
- Discuss the role of leukocytes in the immune system.
- Differentiate between specific and nonspecific defense mechanisms.

I. Life Functions

A. General information

1. All vertebrates have well-developed body systems
2. These systems consist of specialized cells, tissues, and organs to carry out basic properties of life, such as circulation of body fluids, tissue repair, respiration, digestion, movement, waste excretion, reproduction, and growth

B. Body systems

1. Major body systems include the circulatory, respiratory, digestive, neurosensory, musculoskeletal, immune, and reproductive systems
2. The *circulatory system*—a closed network that transports fluids (such as blood and lymph) and other vital substances (such as oxygen and electrolytes) throughout the body—primarily consists of the heart, blood vessels, and blood
3. The *respiratory system* controls the exchange of gases (oxygen and carbon dioxide) necessary to convert glucose into energy used by the cells

4. The *digestive system* breaks down food into molecules that can be absorbed more readily by the body
5. The *nervous system*—which encompasses the brain, sensory organs, and an extensive network of nerves—coordinates sensory input with motor output
6. The *musculoskeletal system* provides internal support and allows for movement
7. The *immune system* is an internal defense mechanism that guards against invading organisms
8. The *reproductive system* is discussed in Chapter 4, "Reproduction and Growth"

II. Circulatory System

A. General information

1. The circulatory system (also called the cardiovascular system) is a self-contained (closed) network that transports fluids (such as blood and lymph) and other vital substances (such as oxygen and electrolytes) throughout the body
2. Primarily composed of the heart, vessels, and blood, the circulatory system along with the lymphatic system exchanges fluid and other substances across capillary membranes
3. Different classes of vertebrates have special adaptations, such as modified heart structures and alternate blood flow routes, that allow for more efficient circulation

B. The heart

1. A muscular sac that pumps blood throughout the circulatory system, the heart consists of one or more atria and one or more ventricles
   a. *Atria* are chambers that receive blood returning to the heart
   b. *Ventricles* are chambers that pump blood from the heart
2. The heart cycle—the sequence of events that occurs during each heartbeat—has two phases: systole and diastole
   a. In *systole,* the ventricles contract and blood is pumped into the arteries
   b. In *diastole,* the ventricles relax and blood from the atria fills the ventricles
3. Valves in the heart prevent the blood from flowing backward when the ventricles contract
4. The human heart has two sets of valves—atrioventricular and semilunar
   a. *Atrioventricular valves* are situated between the atria and the ventricles
      (1) The tricuspid valve separates the right atrium from the right ventricle
      (2) The bicuspid (or mitral) valve separates the left atrium from the left ventricle
   b. *Semilunar valves* are located at the heart's two exit sites
      (1) The pulmonic valve separates the right ventricle from the pulmonary artery
      (2) The aortic valve separates the left ventricle from the aorta
5. The *heart rate* (also called the *pulse*) is the number of heartbeats per minute
   a. In mammals, an inverse relationship exists between body size and pulse—that is, the larger the size, the lower the pulse (for example, elephants

have an average heart rate of 25 beats/minute; shrews, an average heart rate of 600 beats/minute)

b. The relationship between pulse and size is directly related to the surface-to-volume ratio
   (1) The surface-to-volume ratio of smaller animals is greater than that of larger animals; smaller animals, therefore, lose more heat to the surrounding environment
   (2) To compensate for the increased loss of body heat, smaller animals need a higher metabolic rate to maintain a constant body temperature
   (3) A higher metabolic rate leads to a higher oxygen demand by the tissues; blood must be pumped to the tissues at a faster rate to meet this demand

6. *Cardiac output* is the volume of blood per minute pumped from the left ventricle into the systemic circulation; it is influenced by the heart rate and *stroke volume*—the amount of blood pumped by the left ventricle each time it contracts
7. Cardiac muscle cells are self-excitable (meaning they can contract without any signal from the nervous system); despite this intrinsic ability to beat by themselves, cardiac muscle cells must be coordinated with one another for the heart as a whole to function properly
   a. The rate of contraction is set by a specialized region called the *sinoatrial (SA) node,* located in the posterior wall of the right atrium; the SA node is the pacemaker of the heart
   b. Nerves, hormones, body temperature, and exercise influence the SA node and affect the rate of contraction
      (1) One set of nerve fibers (from the sympathetic nervous system) speeds up the SA node; another set (from the parasympathetic nervous system) slows the SA node; the heart rate is a compromise between the effects of these two opposing sets of nerves
      (2) Epinephrine, a hormone secreted by the adrenal glands, can speed up the SA node
      (3) Increased body temperature also can speed up the SA node; an increase of one degree increases the heart rate by 10 to 20 beats/minute
      (4) Exercise also speeds up the SA node, increasing the rate of contraction

C. Blood vessels

1. A network of closed tubes that transport blood throughout the body, blood vessels include arteries, arterioles, capillaries, venules, and veins
2. *Arteries* carry blood away from the heart to organs throughout the body
3. *Arterioles* are branches of arteries within the organs
4. *Capillaries* are microscopic blood vessels that penetrate tissues; their walls consist of a single layer of endothelial cells across which blood and interstitial fluid are exchanged
5. *Venules* are vessels in which capillaries converge
6. *Veins* return blood to the heart; their valves prevent blood from flowing back toward the capillaries

D. Blood

1. Blood consists of a liquid matrix —*plasma*—in which formed elements (red blood cells [RBCs], white blood cells [WBCs], and platelets) are suspended; the average human body contains about 4 to 6 liters of blood
2. Plasma is composed mainly of water (about 90% of the total volume is water) along with dissolved solutes, including electrolytes, proteins, and various transitory substances (such as nutrients, metabolic wastes, respiratory gases, and hormones)
3. The combined concentration of the dissolved solutes determines, to a large extent, the osmotic balance between the blood and the interstitial fluid (fluid outside the blood vessels)
   a. *Electrolytes* are inorganic salts in the form of dissolved ions—for example, sodium, potassium, and calcium
      (1) The kidney regulates the concentration of plasma electrolytes, keeping them in the precise concentrations required by the demands of the body
      (2) Some ions help to buffer the blood from excessive acidity or alkalinity; in humans, they help maintain a blood pH of 7.4
   b. Proteins collectively act to buffer the blood and to determine the osmotic strength of blood; many proteins also have specific functions
      (1) Some proteins are carriers for insoluble substances; for example, thyroid-binding globulin, a special protein, binds with thyroid hormone to help transport the hormone to its target organ
      (2) Some (such as immunoglobulins) combat foreign agents that invade the body
      (3) Others (such as fibrinogen) are clotting factors that plug leaks in damaged blood vessels
4. The blood's formed elements arise from a single source within the bone marrow —a stem cell that differentiates to produce RBCs, WBCs, and platelets
5. RBCs, also called *erythrocytes,* are primarily responsible for transporting oxygen throughout the body
   a. Typically nucleated cells, RBCs have no mitochondria and generate adenosine triphosphate (ATP) by anaerobic metabolism (only mammalian RBCs have no nuclei); their production in the bone marrow is stimulated by erythropoietin, a hormone secreted by the kidney
   b. As RBCs pass through the capillaries of an animal's lungs or gills, oxygen diffuses into the cells and ***hemoglobin*** (a protein containing iron) binds to them; this process is reversed in the capillaries of other organs —the hemoglobin releases oxygen, which diffuses into the surrounding tissues
   c. After circulating for about 120 days, RBCs are destroyed by phagocytic cells located mainly in the liver and spleen
6. WBCs, also called *leukocytes,* primarily defend the body against invading organisms
   a. They form in bone marrow and mature in lymphoid organs (such as the spleen, thymus, tonsils, adenoids, and lymph nodes)
   b. WBCs defend the body by phagocytic action (engulfing foreign particles) or by producing antibodies, which neutralize foreign particles

c. Most WBCs can leave blood vessels and enter the interstitial fluid, where they attack invading organisms

7. Platelets, also called *thrombocytes,* primarily function in blood clotting; they are not really cells, but the pinched-off fragments of large cells originally located in the bone marrow

E. Lymphatic tissue

1. The *lymphatic system*—a network of lymph (a fluid having the same constituents as interstitial fluid), vessels, and nodes—collects and returns blood and proteins lost by the circulatory system into surrounding tissue
   a. The exchange of blood and interstitial fluid across the capillary membrane results in a net loss of fluid (about 3 liters per day) from the capillaries into the tissues; some leakage of blood proteins into the surrounding tissue also occurs
   b. Fluid enters the lymphatic system by diffusing into tiny lymph capillaries situated among the capillaries of the circulatory system
   c. The lymphatic system collects the lost fluid and protein and drains the fluid back into the circulatory system at two locations near the shoulders
2. Like veins, lymph vessels have valves to prevent the backward flow of fluid
3. Along the lymph vessels are specialized tissues called *lymph nodes*
   a. Lymph nodes primarily filter bacteria, viruses, and other foreign matter out of lymph
   b. Because they contain populations of maturing WBCs, lymph nodes are common sites where infection is fought; swollen lymph nodes may be a sign of infection
4. Whenever interstitial fluid accumulates instead of being returned to the blood via the lymphatic system, tissues become swollen with fluid, creating a condition known as *edema*

F. Blood-clotting mechanisms

1. Blood clotting—the process by which blood vessel damage is repaired—involves several steps
   a. When a blood vessel wall is damaged, a portion of the wall containing collagen fibers is pushed into the inside of the vessel (the lumen)
   b. Circulating platelets within the lumen adhere to the damaged wall and release a substance that makes nearby platelets sticky
   c. The "sticky" platelets clump together to form a plug, which is sufficient to seal minor damage to the blood vessel
2. When blood vessel damage is severe, the platelet plug is reinforced by a clot of ***fibrin***
   a. To form a fibrin clot, the clumped platelets release clotting factors that combine with clotting factors circulating in the plasma (calcium and vitamin K must also be present in the plasma for this step to occur)
   b. The mixing together of clotting factors forms an activator, which converts the plasma protein prothrombin into its active form, thrombin
   c. Thrombin promotes the conversion of fibrinogen to fibrin
   d. The fibrin threads become interwoven into a mesh that traps blood cells and seals the injured blood vessel wall

3. The wound is permanently healed by the eventual regeneration of the vessel wall
4. Anticlotting factors normally present in the blood prevent clotting from occurring in the absence of injury
5. Sometimes a clot, called a *thrombus,* forms when a blood vessel has not been damaged; if large enough, a thrombus can block a coronary artery, triggering a heart attack
6. An *embolus* is a thrombus that breaks off from the blood vessel wall and moves through the bloodstream
   a. If large enough, an embolus can become lodged in an artery
   b. An embolus that remains in an artery can damage the heart or, if lodged in an artery supplying blood to the brain, can lead to a stroke

G. Circulatory adaptations

1. Various adaptations of the general circulatory scheme can be found among the vertebrate classes
2. Circulatory adaptations include differences in heart structure and the circulation of blood
3. Fish have a two-chambered heart (one atrium and one ventricle)
   a. Blood is pumped from the ventricle to the gills, where gases are exchanged across a capillary network
   b. The oxygenated blood enters larger blood vessels that carry it to other parts of the body; from these areas, the blood enters another set of capillaries before returning to the atrium
   c. A fish's blood passes through two sets of capillaries in a single circuit
4. Amphibians have a three-chambered heart (two atria and one ventricle)
   a. Blood leaving the ventricle enters a forked artery
      (1) One branch of the artery (the pulmonary circuit) sends blood to the lungs and skin where it becomes oxygenated
      (2) The other branch (the systemic circuit) sends blood to all the other organs
   b. Each circuit has its own set of capillaries
   c. Oxygenated blood returning from the pulmonary circuit enters the left atrium
   d. Deoxygenated blood returning from the systemic circuit enters the right atrium
   e. Although some mixing of deoxygenated and oxygenated blood occurs in the single ventricle, a ridge within the ventricle diverts most of the oxygenated blood from the left atrium to the systemic circuit and most of the deoxygenated blood from the right atrium to the pulmonary circuit
   f. The amphibian's double-circuit circulation has an advantage over the fish's single circuit in that it ensures more vigorous pumping of oxygenated blood to the vital organs
5. Reptiles have a three-chambered heart (two atria and one ventricle) and a double-circuit system; they also have a septum that partially divides the ventricle, resulting in less mixing of deoxygenated and oxygenated blood
6. Birds and mammals have a four-chambered heart (two atria and two ventricles) and a double-circuit system
   a. The two completely separated ventricles keep the deoxygenated blood fully separated from the oxygenated blood

b. Keeping deoxygenated and oxygenated blood apart is an important adaptation for endothermic animals; efficient delivery of oxygen to tissues is necessary to fuel the increased metabolism needed to maintain a constant body temperature

III. Respiratory System

A. General information

1. Respiration is a cellular metabolic process that liberates energy from glucose
2. In such aerobic organisms as vertebrates, respiration requires the exchange of gases (the input of oxygen and the removal of carbon dioxide)
3. The principal organs involved in respiration are the gills (in aquatic animals) and the lungs (in terrestrial animals)

B. Gas exchange

1. The circulatory system transports oxygen to the cells and removes carbon dioxide; these gases are then transported to special gas-exchange organs (lungs, gills, and skin) that exchange gases from the animal with those of the environment
2. To meet the demands of the entire body, gas-exchange organs must be able to take in enough oxygen from the external environment and expel all the carbon dioxide produced by the animal
3. Because oxygen and carbon dioxide can only diffuse across a membrane if they are first dissolved in water, gas-exchange organs in vertebrates have a respiratory surface composed of a single layer of moist epithelium
   a. One side of the epithelial layer is in direct contact with the external environment
   b. The other side is in direct contact with the capillaries

C. Gills

1. Gills are evaginations (or outfoldings) found on the surface of aquatic animals whose external environment and respiratory medium is water
2. Although gills are localized in one area of the body (behind the mouth), their total surface area is greater than that of the body
3. The capillaries in the gills are arranged so that the water and blood flow in opposite directions —a process called *countercurrent exchange;* countercurrent exchange increases the efficiency with which oxygen is transferred to the blood
4. Because the gills are completely surrounded by water, the endothelium is kept constantly moist
5. If an aquatic animal is removed from water, the endothelial surface of the gills quickly dries and the animal dies

D. Lungs

1. Lungs are invaginations (or infoldings) of the body surface of terrestrial animals whose external environment and respiratory medium is air
2. As with gills, lungs are localized in one area of the body (in the chest) and their total surface area is greater than that of the body

3. Lungs maintain contact with the external environment through a narrow tube; this arrangement reduces the loss of water by evaporation and helps to keep the endothelium moist
4. Lungs are highly vascularized; they have a dense network of capillaries just beneath the endothelium

E. Ventilatory adaptations

1. Ventilation is any process that increases contact between the respiratory medium (water or air) and the respiratory surface (gill or lung)
2. Different animals have highly adapted organs or body parts that facilitate ventilation
3. Bony fishes ventilate by using an *operculum*—a flap that covers and pulls water across the gill; because oxygen concentration in water is less than that in air, fish need an efficient means of extracting oxygen; by moving water forcefully over the gill surface, fish can increase the rate of oxygen exchange
4. Terrestrial animals ventilate by means of a *diaphragm*—a large sheet of muscle between the thoracic and abdominal cavities that helps with inhalation and exhalation
    a. Contraction of the diaphragm enlarges the thoracic (chest) cavity, lowering the lung pressure and causing inhalation
    b. Relaxation of the diaphragm increases thoracic pressure, helping to expel air and causing exhalation
    c. This type of breathing is called *negative-pressure breathing*
5. In addition to lungs, birds have eight or nine large air sacs that assist with ventilation
    a. The air sacs do not function directly in gas exchange
    b. They increase ventilation by acting as a bellows to keep air moving into the lungs continuously (during exhalation as well as inhalation)

F. Respiration in mammals

1. All mammals respirate through lungs—large, layered sac-shaped organs located within the thoracic cavity
    a. The inner layer of the sac adheres to the outside of the lungs
    b. The outer layer is attached to the body wall
    c. Between the two layers is a fluid-filled space, which creates a surface tension that allows the inner and outer layers to slide smoothly past each other without pulling apart
2. Before air reaches the lungs, it must be inhaled into the mammal's body
    a. Air enters the body through the nostrils, where it is filtered by hairs, then warmed and humidified
    b. From the nostrils, air proceeds to the *pharynx,* an area of the throat comprising the larynx and esophagus where air and food cross
        (1) Air enters the *larynx,* a rigid structure composed of cartilage that contains the vocal cords
        (2) Food enters the esophagus; food is prevented from entering the larynx by swallowing—a mechanism that pushes the glottis against the epiglottis in such a way that it completely closes off the larynx

c. From the larynx, air passes to the *trachea* (windpipe), another rigid structure composed of rings of cartilage

d. The trachea divides into two smaller tubes called *bronchi,* each of which ultimately leads to a lung

(1) Air enters the bronchi and moves into a series of progressively smaller tubes called *bronchioles*

(2) The smallest bronchioles end in multilobed air sacs called *alveoli,* where gas exchange takes place

(3) The lungs possess millions of alveoli, each one surrounded by a mesh of capillaries

3. Breathing is controlled by a breathing center located within the medulla of the brain

a. Responding to both neural and chemical signals, the breathing center adjusts the rate of breathing to meet the body's changing demands

b. At rest, the breathing center sends nerve impulses to the rib muscles or diaphragm, stimulating these muscles to contract 10 to 14 times per minute

c. If the lung is overstretched, stretch receptors in the lung tissue send messages that inhibit the breathing center

4. The breathing center monitors blood pH levels

a. Carbon dioxide acts as an acid in the bloodstream; if the level of carbon dioxide increases (as in the case of hypoventilation), the pH level of the blood decreases, signaling the breathing center to increase the rate and depth of breathing

b. The breathing center is less sensitive to changes in oxygen concentration than to those in carbon dioxide concentration

c. Oxygen sensors are located in key arteries

d. If the oxygen level falls dramatically, the sensors in the artery send an alarm to the breathing center, signaling it to increase the rate of breathing

IV. Digestive System

A. General information

1. Digestion —the process of breaking food down into molecules small enough for the body to absorb —involves the chemical action of enzymes that breaks the bonds of macromolecules by the addition of water (hydrolysis)

2. This process can occur inside or outside the cell, depending on the type of animal

a. In the simplest animals (such as sponges), digestion occurs in food vacuoles within the cell —a process called *intracellular digestion*

b. In most other animals, digestion occurs in compartments connected by a passageway with the outside of the body —a process called *extracellular digestion*

(1) Cnidarians and flatworms have a digestive sac with a single opening called a *gastrovascular cavity*

(2) Nematodes, annelids, mollusks, arthropods, echinoderms, and chordates (the branch to which vertebrates belong) have a complete digestive tract, or an *alimentary canal,* with two openings —a mouth at one end and an anus at the other end

3. In animals with a complete digestive tract, food moves along the tract in one direction
   a. Food enters the mouth and is acted on by enzymes in successive stages, typically in a series of specialized compartments; some compartments carry out digestion, whereas others carry out absorption
   b. Undigested wastes are excreted from the anus
   c. Special mechanisms, such as chewing, mechanically fragment the food to better prepare it for chemical digestion
   d. Special accessory glands situated along the tract —including three pairs of salivary glands (parotid, sublingual, and submandibular), the pancreas, the liver, and the gallbladder —secrete digestive juices to break down food
4. The alimentary canal consists of four layers: the mucosa (which lines the lumen), submucosa, smooth muscle, and peritoneum
   a. The *mucosa* is a layer of epithelial tissue containing mucus-secreting cells that help to protect the walls of the digestive tract and provide lubrication
   b. The *submucosa* is a layer of connective tissue containing blood vessels, nerves, and structures that secrete digestive enzymes
   c. *Smooth muscle,* a special type of tissue, produces rhythmic contractions that push food along the digestive tract (a process called *peristalsis*)
      (1) Smooth muscle may become specialized into ring-like valves, called *sphincters,* at various junctures between compartments
      (2) Sphincters regulate the passage of food from one compartment to the next
   d. The *peritoneum* is a layer of connective tissue that attaches to the membrane of the abdominal cavity and supports the abdominal organs
5. The alimentary canal contains several compartments; in humans, these include the oral cavity, pharynx, esophagus, stomach, small intestine, and large intestine

B. Oral cavity

1. The oral cavity is the site where mechanical fragmentation and chemical digestion of food begins; it encompases such structures as the tongue, salivary glands, and teeth
2. Mechanical fragmentation is accomplished by chewing, which makes food easier to swallow and increases the surface area on which enzymes can act
   a. The tongue is used to taste food and manipulate it during chewing
   b. The ball of food formed at this stage is called a *bolus*
3. In swallowing, the tongue pushes the food to the back of the oral cavity and into the pharynx
4. The presence of food in the oral cavity triggers a nervous system reflex that causes the salivary glands to deliver saliva —a slippery fluid secreted in the mouth to aid the breakdown and passage of food —to the oral cavity
5. Saliva consists of four basic components: mucin —a slippery glycoprotein that protects the soft lining of the mouth from abrasion and lubricates the food; buffers, which neutralize acids in the mouth and help to prevent tooth decay; antibacterial agents; and salivary amylase —a digestive enzyme that hydrolyzes starch

C. Pharynx

1. The pharynx, commonly called the throat, connects the oral cavity with the esophagus and the larynx
2. When an animal swallows, the epiglottis closes over the larynx to prevent aspiration of food into the lungs

D. Esophagus

1. The esophagus—a muscular, collapsible tube located behind the trachea—connects the pharynx with the stomach
2. Peristalsis pushes food along the length of this collapsible tube
3. Muscles at the upper and lower ends of the tube serve as sphincters, regulating the passage of food into and out of the esophagus

E. Stomach

1. Located on the left side of the abdominal cavity just below the diaphragm, the stomach serves as a food storage area during the early part of digestion
2. It can stretch to accommodate about 2 liters of food and fluid in humans
3. The stomach is divided into the fundus (upper part), the body (middle part), and the antrum (lower part)
4. The stomach constantly churns, mixing up food contents through the action of its smooth muscles
5. The epithelium of the stomach secretes gastric juice, which has two major components —hydrochloric acid and pepsin
   a. *Hydrochloric acid* is secreted by parietal cells (special epithelial cells found mostly in the fundus)
   b. *Pepsin* is an enzyme originally secreted by chief cells (another type of specialized epithelial cell found mostly in the fundus) in an inactive form called *pepsinogen;* the hydrochloric acid in gastric juice converts pepsinogen to pepsin
6. Hydrochloric acid, which gives gastric juice a pH of about 2.0 (highly acidic), breaks up complex macromolecules found in meat and plant material and destroys bacteria
   a. Large amounts of mucus are secreted by other cells to protect the stomach against injury from hydrochloric acid
   b. Occasionally, hydrochloric acid may erode the wall of the stomach or the small intestine; this erosion is called an *ulcer*
7. Pepsin hydrolyzes proteins into smaller polypeptides
8. The release of gastric juice is regulated by negative feedback
   a. The smell, sight, or taste of food initiates a nerve impulse that travels from the brain to the stomach, triggering the release of gastric juice
   b. As the food is beginning to be digested in the stomach, substances in the food stimulate the antrum to release gastrin; this hormone causes more gastric juice to be produced
   c. When the pH level of the stomach contents becomes too low, the release of gastrin is inhibited and the flow of gastric juice stops

F. Small intestine

1. The longest section of the alimentary canal, the small intestine is divided into the duodenum, the jejunum, and the ileum

2. The *duodenum,* the first segment of the small intestine, is the site where most digestion is completed
   a. Separated from the stomach by the pyloric valve, the duodenum receives chyme from the stomach, pancreatic juice from the pancreas, and bile from the liver and gallbladder; its epithelium secretes a group of enzymes collectively referred to as *intestinal juice*
      (1) Chyme is an acidic liquid secreted by the stomach into the duodenum; it contains gastric juice mixed with partially digested food
      (2) Pancreatic juice enters the duodenum via the pancreatic duct; these secretions include pancreatic amylase, sodium bicarbonate, trypsin, chymotrypsin, carboxypeptidase, and lipase
         (a) *Pancreatic amylase* hydrolyzes starch into maltose, a disaccharide
         (b) *Sodium bicarbonate,* a highly alkaline substance, neutralizes the acidic chyme
         (c) *Trypsin* and *chymotrypsin* break down polypeptides into smaller polypeptides
         (d) *Carboxypeptidase* breaks down small polypeptides into amino acids
         (e) *Lipase* breaks down fats into glycerol, fatty acids, and glycerides
      (3) Bile —a yellowish green fluid secreted by the liver and stored in the gallbladder —enters the duodenum via the common bile duct
         (a) Bile contains bile salts and pigments from the degradation of hemoglobin
         (b) Bile salts coat tiny fat droplets and keep them from coalescing (a process called *emulsification,* which greatly aids fat digestion)
         (c) Bile pigments are eliminated from the body with the feces and give the feces their characteristic color
   b. The intestinal juice secreted by the duodenum includes aminopeptidase and the disaccharidases maltase, lactase, and sucrase
      (1) *Aminopeptidase* has an action similar to that of carboxypeptidase; it breaks down small polypeptides into amino acids
      (2) *Maltase* splits maltose into two molecules of the simple sugar glucose
      (3) *Lactase* hydrolyzes lactose into galactose and glucose
      (4) *Sucrase* hydrolyzes sucrose into fructose and glucose
3. The *jejunum* and the *ileum* have a lining specialized for the absorption of nutrients
   a. The large folds of the lining greatly increase its surface area
   b. Each fold contains a number of ***villi*** (fingerlike projections containing numerous epithelial cells)
   c. Penetrating each villus is a network of capillaries and a lymph vessel called a *lacteal*
   d. Amino acids and sugars pass through the epithelial cells of the villus and enter the capillaries, where they are carried away from the intestine by the bloodstream
   e. Glycerol and fatty acids enter the epithelial cells of the villus and are recombined to form fats
      (1) Some of these fat molecules are bound to specialized proteins and transported as lipoproteins directly into the capillaries

(2) Other fat molecules are coated with other specialized proteins to make tiny globules called *chylomicrons,* which are transported into the lacteals

f. All the capillaries of the villi drain their nutrients into a single large vessel, the hepatic portal vein, which carries the nutrients directly to the liver

4. Three hormones —secretin, cholecystokinin (CCK), and enterogastrone —regulate digestive secretion in the small intestine
   a. *Secretin* signals the pancreas to release bicarbonate; its release is stimulated by the acidic pH level of the chyme as it enters the duodenum
   b. *CCK* causes the gallbladder to release bile and the pancreas to release pancreatic enzymes; its release is stimulated by the presence of amino acids or fatty acids in chyme
   c. *Enterogastrone* inhibits peristalsis in the stomach, which slows the entry of food into the small intestine; its release is stimulated by the presence of fats in chyme

G. Large intestine

1. The large intestine (or colon) connects the ileum of the small intestine with the rectum —an area where feces are stored until elimination
2. The large intestine is divided into three parts: the ascending colon, the transverse colon, and the descending colon
3. At the beginning of the ascending colon is a blind pouch called the *cecum;* the cecum includes a fingerlike extension, the appendix, which is made up of lymphoid tissue
4. As digestive tract wastes move through the colon by peristalsis, the feces become more solid
5. The large intestine reabsorbs water (together, the small intestine and large intestine reabsorb about 90% of the water that originally enters the alimentary canal); when too much or too little water is reabsorbed, the consistency of feces is altered
   a. If peristalsis is too slow, an excess of water is reabsorbed and constipation occurs
   b. If the lining of the colon is irritated by a viral or bacterial infection, less water is reabsorbed and diarrhea results

H. Digestive adaptations

1. Different vertebrates have special adaptations to help digest food
2. These adaptations include the type of dentition (teeth), the length of the digestive tract, and the presence of special fermentation chambers
3. The dentition of an animal reflects the animal's diet
   a. Carnivores have pointed teeth for killing prey and ripping away flesh
   b. Herbivores have teeth with broad-ridged surfaces for grinding
   c. Omnivores (such as humans) have both pointed and broad-ridged teeth
4. The digestive tract of an animal is relative to its body size (for example, the digestive tract of a herbivore or an omnivore is longer than that of a carnivore); the difference in length is necessary because plants are usually more difficult to digest than meat

5. Special fermentation chambers containing symbiotic bacteria and protozoa are found in the alimentary canals of many herbivores
    a. The bacteria and protozoa digest cellulose (which animals themselves cannot digest)
    b. During digestion, these microorganisms release simple sugars and other essential nutrients

V. Nervous System

A. General information

1. The nervous system of vertebrates is divided into a central nervous system and a peripheral nervous system
    a. The *central nervous system,* which comprises the brain and spinal cord, is the division that controls the integration of information
    b. The *peripheral nervous system,* which comprises a network of nerves, controls communication between the central nervous system and the rest of the body
2. The nervous system comprises two types of cells —neurons and supporting cells
    a. The functional units of the nervous system, *neurons* are specialized cells that communicate information between receptors (sensory organs that receive stimuli) and effectors (muscles, glands, or organs that receive information from the brain and effect a change)
        (1) They have a relatively large cell body containing a nucleus, cytoplasm, and organelles
        (2) Neurons have two types of cellular extensions —dendrites and axons
            (a) *Dendrites* transmit signals toward the cell body
            (b) *Axons,* which transmit signals away from the cell body, may be branched, with each branch ending in branchlets whose ends contain synaptic knobs
            (c) The synaptic knobs release *neurotransmitters* (chemical messengers), which diffuse across the narrow gap (***synapse***) between adjacent neurons
    b. *Supporting cells* (also called *glial cells)* assist neurons by providing structural reinforcement, protection, and insulation; three types of supporting cells are found in vertebrates
        (1) *Astrocytes* line the capillaries of the brain and contribute to the blood-brain barrier —a barrier that restricts the passage of most substances into the brain and protects the brain from possible injury
        (2) *Oligodendrocytes* insulate neurons in the central nervous system by forming protective coverings (myelin sheaths)
        (3) *Schwann's cells* are the source of myelin; they are also the only covering for the nodes of Ranvier —small gaps in the nerve fibers

B. Central nervous system

1. The central nervous system primarily consists of the spinal cord and the brain; neurons within this division of the nervous system are called *interneurons*

2. The *spinal cord* is a collection of nervous tissue that runs down the neck and back inside the vertebral column; it receives information from the skin and muscles and transmits motor commands for movement
3. The *brain,* which is located at the top of the spinal cord, is involved in more complex activities, such as perception, coordination of movement, integration of homeostasis and, in humans, intellect and emotions
4. In the central nervous system, axons that are myelinated (covered by a myelin sheath) have a white appearance; those that are not myelinated have a gray appearance
   a. White matter is located in the brain's inner regions; gray matter, the outer regions
   b. The reverse is true in the spinal cord, where white matter is on the outside and gray matter on the inside
5. A set of protective membranes, or meninges, cover the entire central nervous system; from innermost to outermost, the three meninges are the pia mater, the arachnoid, and the dura mater
6. Cerebrospinal fluid fills the ventricles of the brain and the central canal of the spinal cord

C. Peripheral nervous system

1. The peripheral nervous system consists of bundles of neurons that communicate between the central nervous system and the rest of the body
   a. *Sensory neurons,* also called *afferent neurons,* convey signals from sensory receptors to the central nervous system
   b. *Motor neurons,* also called *efferent neurons,* convey signals from the central nervous system to effector cells
2. The peripheral nervous system is divided into two parts: the somatic nervous system and the autonomic nervous system
   a. The *somatic nervous system* is primarily responsible for conveying sensory information
   b. The *autonomic nervous system* is primarily responsible for controlling motor information; it regulates the internal environment by controlling the smooth and cardiac muscles and the organs of the cardiovascular, endocrine, excretory, and GI systems
3. The autonomic system is divided into a sympathetic nervous system and a parasympathetic nervous system
   a. When stimulated, the *sympathetic nervous system* prepares an animal for fight or flight by accelerating the heart rate; increasing the metabolic rate; constricting the peripheral capillaries so that blood is shunted to the heart, lungs, and brain; and dilating the bronchioles of the lungs so that more air can be taken in
   b. The *parasympathetic nervous system* enhances activities to conserve energy for the animal, such as slowing the heart rate and stimulating digestion
4. When the sympathetic and parasympathetic systems innervate the same organ, they typically have opposite (antagonistic) effects

D. Transmission of nerve impulses

1. Impulses—signals transmitted along a neuron (from dendrite to axon)—are electric currents generated by ion movements across the plasma membrane of the cell
   a. A nontransmitting neuron has potassium ions on the inside of the plasma membrane and sodium ions on the outside of the membrane
   b. An ATP-driven sodium pump, located in the plasma membrane, actively transports sodium ions to the outside and potassium ions to the inside
   c. The difference in electrical charges across the membrane created by the separation of these two types of ions can be measured as a type of voltage, or *membrane potential*
      (1) The voltage of a nontransmitting neuron is called the *resting potential* and is equal to about 70 millivolts
      (2) When the neuron achieves maximum separation of the electrical charges across the membrane, it is polarized
   d. In addition to the ATP-driven sodium pump, the neuron's membrane also has a series of channels, or gates, as part of its microstructure
      (1) Neurons have two types of gates—calcium-specific channels and sodium-specific channels, which are made of proteins embedded in the membrane
      (2) When opened, these channels allow ions to pass through
      (3) In the nontransmitting neuron, all channels are closed
2. A stimulus is any environmental factor that can alter the permeability of the cell membrane and decrease the maximum voltage (resting potential)
   a. It can be mechanical, such as pressure or temperature, or chemical
   b. A stimulus depolarizes the cell membrane by allowing sodium to diffuse across the membrane into the cell
   c. Leakage of sodium across the cell membrane partially depolarizes the cell, creating a *graded potential*
   d. Graded potentials are proportional to the strength of the stimulus—the stronger the stimulus, the more sodium leaks across the membrane
3. If the depolarization caused by a graded potential is large enough (reaches a critical threshold), the neuron's gates open and sodium rushes in, completely depolarizing the cell and generating an ***action potential;*** if the graded potential is not large enough to reach a critical threshold, nothing further happens
   a. An action potential is a rapid, reversible, complete depolarization of the cell
   b. It is an all-or-nothing event that changes the voltage from the resting potential of 70 millivolts to the action potential of 35 millivolts
4. After the action potential moves down the cell membrane, the membrane is completely depolarized; it begins to repolarize by using the ATP-driven sodium pump to push the sodium outside and pull the potassium in
5. During repolarization, no stimulus can affect the cell and the cell is said to be *refractory*
6. When the action potential reaches the end of the axon, it depolarizes the membrane of the synaptic vesicle, which causes the vesicle to release a *neurotransmitter*—a chemical stimulus that diffuses across the synapse, generating a graded potential in the dendrites of the next neuron

7. If the graded potential is strong enough, it opens the sodium gates of the neighboring neuron's membrane and generates an action potential, repeating the entire depolarization-polarization process
8. Neurotransmitters do not always excite a neuron; in some cases, they may inhibit the cell by causing hyperpolarization of the cell's membrane
9. Common neurotransmitters include acetylcholine, biogenic amines, amino acids, and neuropeptides
   a. *Acetylcholine* is one of the most common neurotransmitters
      (1) It usually is found in the synapse between a motor neuron and skeletal muscle cell, where it triggers contraction of the cell
      (2) Its action on heart muscle inhibits the heart rate
   b. *Biogenic amines* include epinephrine, norepinephrine, dopamine, and serotonin
      (1) These neurotransmitters usually are found in the central nervous system (although norepinephrine is found in the peripheral nervous system)
      (2) Their action in the brain can affect mood, sleep, attention span, and learning
   c. The three *amino acid neurotransmitters*—glycine, glutamate, and gamma aminobutyric acid (GABA)—function in the central nervous system; GABA is a major inhibitory neurotransmitter in the brain
   d. *Neuropeptides* are short-chain amino acids
      (1) They include the endorphins and the enkephalins
      (2) These two groups of neurotransmitters function as analgesics, decreasing the perception of pain by the central nervous system

E. The human brain

1. In all vertebrates, the brain began as a series of three bulges—rhombencephalon (hindbrain), mesencephalon (midbrain), and prosencephalon (forebrain)—at the anterior end of the spinal cord; as vertebrates evolved into more complex forms, these three regions became further subdivided and more specialized
2. In humans, these three regions encompass several subdivisions
   a. The hindbrain is subdivided into the medulla oblongata, pons, and cerebellum
   b. The midbrain is subdivided into the superior and inferior colliculi and the reticular formation
   c. The forebrain is divided into the diencephalon and the telencephalon
      (1) The diencephalon contains the thalamus and hypothalamus
      (2) The telencephalon contains the basal ganglia, limbic system, and cerebral cortex
3. The *medulla oblongata* and *pons* control many autonomic functions, including breathing, heart and blood vessel activity, swallowing, vomiting, and digestion
   a. These two regions of the hindbrain play a major role in conducting information throughout the body; all the sensory and motor neurons passing to and from higher brain regions must pass through the medulla and the pons
   b. Most motor fibers from the midbrain and forebrain cross over at the medulla (for example, a motor neuron that originates on the left side of the brain crosses over at the medulla to innervate muscles on the right side)

4. The *cerebellum* coordinates movement and balance
5. The *superior* and *inferior colliculi* are part of the visual and auditory systems; the superior colliculi coordinates visual reflexes, such as blinking; the inferior colliculi is the site where all the fibers involved in hearing either terminate or pass through to the ears
6. The *reticular formation* regulates states of arousal; it determines which sensory input reaches the cerebral cortex
7. The *thalamus* works with the reticular formation, filtering and conveying information to the cerebral cortex
8. The *hypothalamus* is critical to the regulation of homeostasis
    a. It is the source of posterior pituitary hormones and anterior pituitary releasing factor
    b. It also regulates hunger and thirst and contains a pleasure center
9. The *basal ganglia* lie just below the cerebral cortex and function in motor coordination
10. The *limbic system,* located in the lower part of the forebrain, selects which emotional and behavioral responses reach the cerebral cortex
11. The *cerebral cortex* is the largest, most complex, and most highly evolved part of the human brain
    a. It is bilaterally symmetrical, with a right and a left hemisphere that are connected to each other by fibers known as the *corpus callosum*
    b. Its highly convoluted shape greatly increases its total surface area
    c. The cerebral cortex is divided into four lobes —frontal, temporal, parietal, and occipital
        (1) The highest cognitive and motor functions originate in the *frontal* lobe; the left frontal lobe contains a speech area (Broca's area) that controls the movement of muscles used for articulation and the retreival necessary for deciding what to say)
        (2) The *temporal* lobe coordinates and integrates olfaction (smell) and hearing
        (3) The *parietal* lobe functions in somatic associations (perceptions or meanings); it contains an area where sensory input is interpreted and where information required for understanding speech is processed
        (4) Visual association occurs in the *occipital* lobe

VI. Special Sensory Systems

A. General information

1. In vertebrates, vision, hearing, and balance are controlled by special sensory organs that detect energy waves and change them into nerve impulses that can be transmitted to the brain
    a. Vertebrates see by converting light energy into nerve impulses that are transmitted to the brain
    b. They hear by converting the energy of pressure waves traveling through the air into nerve impulses that are transmitted to the brain
    c. They maintain balance and equilibrium by constantly monitoring their position with respect to gravity

2. The principal sensory organs controlling vision, hearing, and balance are the eyes and ears

B. Vision

1. The vertebrate eye comprises three layers—the sclera, the choroid, and the retina
   a. The *sclera*—the outermost layer—is made up of connective tissue; at the front of the eye, the sclera becomes the transparent cornea
   b. The *choroid*—the middle layer—is rich in blood vessels and nourishes the other two layers
      (1) At the front of the eye, the choroid becomes the *iris,* a pigmented section that gives the eye its characteristic color
      (2) In the center of the iris is an opening called the *pupil,* through which light enters the eye
      (3) The choroid also contains a muscular ciliary body attached to the lens
         (a) Contraction of the muscles in the ciliary body alters the shape of the lens, enabling it to focus an image
         (b) Contraction and relaxation of the ciliary body also alters the size of the pupil, thereby controlling the amount of light that enters the eye
   c. The *retina*—the innermost layer—is made up of special photoreceptor cells called *rods* and *cones*
      (1) Rods and cones convey visual information to the brain via the optic nerve
      (2) The point at which the optic nerve attaches to the eye is called the *optic disc*
      (3) The *fovea* is an area of the retina that acts as the center of the visual field
2. The *lens* divides the eye into two cavities, each of which is filled with a special fluid that helps the eye to focus
   a. The cavity between the lens and the cornea is filled with *aqueous humor*
      (1) This fluid is produced continuously by the ciliary body and drains off through a special system of ducts
      (2) If the ducts are blocked, glaucoma can result; this condition is characterized by increased pressure in the cavity that can eventually lead to blindness
   b. The cavity behind the lens is much larger and filled with a jelly-like substance called *vitreous humor*

C. The human eye

1. The human eye contains 125 million rods and 6 million cones; these are modified neurons that together constitute 70% of all the sensory receptors in the body
2. Rods are found in greatest density at the lateral parts of the visual field and are absent in the fovea
   a. They are very sensitive to light but do not distinguish color
   b. They are needed for night vision

3. Cones are less sensitive to light and can distinguish colors
4. Both rods and cones have a specialized outer segment consisting of stacks of folded membranes; this area contains the visual pigments
   a. The visual pigment consists of retinal (derived from vitamin A) and the protein opsin
   b. Rods and cones have their own type of opsin
      (1) The combination of retinal and opsin in rods is called *rhodopsin*
      (2) The same combination in cones is called *photopsin*
   c. Cones are divided into three subsets —red, green, and blue —based on the type of opsin that they contain
      (1) Seeing the full range of colors results from the mixing of signals from the three types of cones; for example, when red and blue cones are stimulated, the eye may see yellow
      (2) Color blindness, an inherited, sex-linked trait, is caused by a deficiency in one or more of the three types of cones

D. The visual response

1. Vision results from changes in the visual pigment (rhodopsin or photopsin) caused by interaction with light
2. The visual response to light involves a series of steps
   a. If the light that strikes a rod or cone is strong enough and the opsin in that rod or cone is specific to the wavelength of the light, the visual pigment absorbs the light
   b. The absorbed light causes the retina to change shape and to dissociate from the opsin; this photochemical reaction is called *bleaching*
   c. When light does not strike a rod or cone, enzymes convert the visual pigment back to its original form; in the presence of light, the visual pigments remain bleached and are unable to respond to subsequent stimulation by light
   d. To complete the visual response, the photochemical reaction to light occurring in the rods or cones must be transmitted to the visual cortex of the brain

E. Vision transmission

1. When the retina changes shape, it creates a series of metabolic events that alter membrane potentials
2. The altered membrane potential is a graded potential; if strong enough, this graded potential stimulates an action potential in the bipolar cell adjacent to the rod or cone
   a. This action potential is transmitted from one neuron to the next
   b. All action potentials feed into the optic nerve
3. The optic nerve transmits information from the retina to the visual cortex, where the information is interpreted to create an image
   a. Each eye has its own optic nerve
   b. Some fibers within the optic nerves cross at the base of the brain; thus some signals originating in the right eye are interpreted by the left side of the brain and vice versa

F. Hearing and balance

1. An animal hears by converting the energy of pressure waves traveling through the air into nerve impulses that are transmitted to the brain
2. An animal maintains balance and equilibrium by constantly monitoring its position with respect to gravity
3. Both hearing and balance involve the manipulation of mechanoreceptors containing hair cells; when the hairs are bent by settling particles or moving fluid, they trigger an action potential that initiates a nerve impulse transmission
4. Most fish and aquatic amphibians have a lateral line system that runs along both sides of the body
    a. This system enables the animals to monitor water currents, detect pressure waves produced by moving objects, and perceive low-frequency sounds
    b. Water moving through this system stimulates hair cells contained in special structures called *neuromasts*
5. Fish also have inner ears located near their brain that enable them to hear; the ears do not open to the outside of the body and have no eardrums
6. The mammalian ear controls hearing and balance; it consists of an outer ear, a middle ear, and an inner ear
    a. The *outer ear* consists of the pinna and the auditory canal, both of which collect and channel sound waves
    b. The *middle ear* consists of the tympanic membrane and three ossicles—the malleus (hammer), incus (anvil) and stapes (stirrup)
        (1) The eustachian tube connects the middle ear with the pharynx
        (2) It equalizes pressure between the middle ear and the atmosphere
    c. The *inner ear* consists of the oval window (a membrane beneath the stapes), the cochlea (a fluid-filled structure that contains a special membrane with hair cells called the organ of Corti), the vestibule (a structure that consists of two chambers—the utricle and the saccule), and the semicircular canals
        (1) The vestibule and the semicircular canals are filled with a gelatinous substance that contains numerous small calcium carbonate particles called *otoliths*
        (2) The vestibule and the semicircular canals also contain hair cells that project into the gelatinous mass
7. In mammals, hearing begins with the vibration of the tympanic membrane; this membrane vibrates with the same frequency as that of the sound from the surrounding air
    a. The ossicles amplify and transmit the sound waves
    b. The mechanical vibration of the stapes on the oval window of the cochlea produces pressure waves in the fluid within the cochlea
    c. The cochlea transduces (converts) the energy of the vibrating fluid into action potentials; the hair cells within the organ of Corti are depolarized by the action of vibration
    d. The action potentials are transmitted via the auditory nerve to the brain
    e. The original sound wave eventually strikes the round window (located beneath the oval window) and dissipates
8. Balance and equilibrium in mammals are controlled by the vestibule and the semicircular canals, which contain special hair cells that monitor the body's position

a. A change in gravity alters the force exerted on these hair cells
b. This in turn alters the messages sent by these cells to the brain

VII. Musculoskeletal System

A. General information

1. Vertebrates have an internal supporting framework, or endoskeleton
2. In mammals, this skeleton contains more than 200 bones; some are fused together, whereas others are connected at joints by ligaments
3. Skeletal muscle, which is attached to the bones, enables an animal to move

B. Endoskeleton

1. The vertebrate endoskeleton is divided into an axial skeleton and an appendicular skeleton
    a. The *axial skeleton* consists of the skull, vertebral column, and rib cage
    b. The *appendicular skeleton* consists of the limb bones (appendages such as arms, legs, or wings) and the pectoral and pelvic girdles, which anchor the appendages to the axial skeleton
2. The bones of the endoskeleton act as levers when the muscles attached to them contract

C. Skeletal muscle

1. Skeletal muscle is composed of a bundle of long fibers that runs the length of the muscle
    a. Each fiber is made up of a bundle of myofibrils
    b. Each myofibril is made up of two types of myofilaments —thin filaments and thick filaments
        (1) *Thin filaments* consist of two strands of the protein actin and one or more regulatory proteins wrapped around the actin
        (2) Thick filaments are composed of the protein myosin
2. Skeletal muscle is called *striated muscle* because the regular, repeating arrangement of the myofilaments creates a striated appearance (alternating dark and light bands)
    a. Each repeating unit of myofilaments is called a ***sarcomere***
    b. The sarcomere has several landmarks —the Z line, the I band, the A band, and the H zone
        (1) The *Z line* is the point at which the thin filaments are attached; the distance between two Z lines equals one sarcomere unit
        (2) The *I band* is an area where only thin filaments occur
        (3) The *A band* is an area where thin and thick filaments overlap
        (4) The *H zone* is an area in the center of the A band where only thick filaments occur

D. Movement

1. An animal moves by contracting its muscles
2. The mechanism of muscle contraction is described by the *sliding filament model*
    a. Contraction occurs when thin filaments slide (more precisely, ratchet) across thick filaments

b. When contraction occurs, the thick filaments pull the Z lines together and shorten the sarcomere
c. Also during contraction, the I bands shorten but do not change in length
3. The basis for the sliding of filaments is the molecular interaction of ***actin*** and ***myosin***
a. Myosin molecules contain a special area (the myosin head) with an affinity for actin molecules; it forms bridges with the actin at specific points
b. Attached to the myosin head is ADP and a phosphate; release of these substances from the myosin head (the binding of calcium to troponin acts as a stimulus for this to occur), causes the cross-bridge to bend; this is known as the *power stroke* because of the force necessary to bend the cross-bridge and make the filaments slide past one another
c. Energy stored in ATP releases the myosin head from the actin filament; hydrolysis of ATP results in the formation of ADP and phosphate, which attach to the myosin head
d. The myosin head then reattaches to its original position on the actin
4. The resting actin molecule is in contact with two regulatory proteins that prevent the myofilament from contracting
a. One protein, tropomyosin, is a long rodlike molecule that is intertwined along the length of the actin
b. The other, troponin, sits on the myosin binding sites
5. For contraction to occur, these two proteins must be altered to allow the myosin head to bind to the actin at the myosin binding sites; this occurs when calcium binds to troponin
6. The interaction between calcium and troponin alters the shape of the filament, exposing the myosin binding site; calcium, therefore, is crucial to muscle contraction; if no calcium is available, contraction cannot occur
7. The calcium concentration in the cytoplasm muscle cell is regulated by specialized endoplasmic reticulum called ***sarcoplasmic reticulum;*** the membrane of the reticulum actively transports calcium from the cytoplasm into the spaces within the sarcoplasmic reticulum, where it is stored

VIII. Immune System

A. General information

1. The immune system is a biochemical complex that protects an animal against pathogenic organisms and other foreign bodies
2. It incorporates both the humoral immune response and the cell-mediated response
a. The *humoral immune response* produces antibodies to react with specific antigens
(1) An *antibody* is a specific defensive protein produced by the immune system
(2) An ***antigen*** is any foreign substance that invades the body
b. The *cell-mediated response* uses T cells to mobilize tissue macrophages in the presence of a foreign body
3. WBCs, or *leukocytes,* enable the immune system to carry out its functions
4. All WBCs are derived from a precursor cell in the bone marrow

5. Leukocytes are divided into two populations based on the shape of their nuclei — polymorphonuclear leukocytes (also called granulocytes) and mononuclear leukocytes

B. Polymorphonuclear leukocytes

1. Polymorphonuclear leukocytes include neutrophils, basophils, and eosinophils
2. *Neutrophils* (also called *microphages*) are the most numerous of the polymorphonuclear cells
   a. These rapidly migrating cells appear in areas of infection or tissue damage, where they engulf invading organisms
   b. They form the first wave of cellular attack against infection and are characteristically seen in acute inflammation
   c. Their cytoplasm contains pink-staining granules filled with hydrolytic enzymes that are capable of digesting foreign material
3. *Basophils* are also present at the site of an infection
   a. They produce the symptoms associated with early inflammation
   b. Their cytoplasm contains blue-staining granules filled with heparin, histamine, serotonin, hydrolytic enzymes, and a substance that can be metabolized to form prostaglandin
4. *Eosinophils* appear in large numbers in tissues infected with parasites
   a. They appear to play a role in limiting or modulating inflammation
   b. Their cytoplasm contains red-staining granules filled with hydrolytic enzymes

C. Mononuclear leukocytes

1. Mononuclear leukocytes comprise the monocytes (also called macrophages) and the lymphocytes
2. *Macrophages* are capable of phagocytosis
3. *Lymphocytes* include T cells, B cells, and natural killer cells
   a. The "T" in T cells stands for "thymus"; these cells leave the bone marrow and migrate to the thymus to complete their maturation
      (1) T cells function in cell-mediated immunity; they are the master regulators and effectors of the immune system
      (2) Helper T cells boost the immune response, whereas suppressor T cells repress it
      (3) Cytotoxic T cells kill invading cells directly by releasing a substance that brings about the lysis (destruction) of the cell
   b. The "B" in B cells stands for "bursa"; in birds, these cells mature in a lymphoid organ called the bursa of Fabricius (for some time it was thought that humans had a similar, yet undiscovered organ; the consensus today is that humans have no equivalent organ and that human B cells mature in the bone marrow)
      (1) B cells function in humoral immunity
      (2) When a B cell is stimulated, it can differentiate into a plasma cell that produces antibodies
   c. *Natural killer cells* attack the invading microorganism indirectly by destroying the body's own cells that have become infected

D. Nonspecific defense mechanisms

1. The skin and mucous membranes serve as the first line of defense against invading pathogens
   a. The skin provides a physical barrier to microorganisms reinforced by chemical defenses
      (1) Secretions from sweat and oil glands give human skin a pH level of 3.5, which is acidic enough to kill most bacteria
      (2) Perspiration also contains the enzyme lysozyme, which attacks the cell walls of many bacteria
   b. The mucus that lines the respiratory tract traps microorganisms; cilia move the mucus and any trapped particles out of the body
2. Phagocytes, including the neutrophils and macrophages, engulf foreign particles; these cells wander through the interstitial fluid engulfing bacteria, viruses, and the debris of damaged cells
3. Inflammation —a protective response by body tissue to injury or irritation —can be localized or systemic
   a. In a *localized* response, small blood vessels near the injury dilate and their permeability changes to allow for easier movement of phagocytic cells from the blood to the interstitial tissue
      (1) Redness, swelling, and warmth are signs of a localized infection
      (2) Histamine, which is released by the damaged cells at the site of injury, plays an important role in this response; it causes dilation of the blood vessel and changes the vessel's permeability
   b. In a *systemic* response, the number of WBCs circulating throughout the body is increased; fever is a sign of a systemic infection
4. Antimicrobial proteins (interferons and complement) help to fight invading microorganisms
   a. Several types of ***interferons*** exist, each produced by a different cell; these proteins interfere with the growth and reproduction of viruses
   b. *Complement* is a group of about 20 proteins that circulate in an inactive form
      (1) When activated by the presence of a foreign substance, these proteins amplify the inflammatory response by stimulating histamine release and attracting phagocytes
      (2) Some complement proteins coat the invading microorganisms; this process, called *opsonization,* facilitates the engulfment of pathogens by the phagocytes

E. Specific defense mechanisms

1. Specific defense mechanisms include humoral immunity and cell-mediated immunity
   a. *Humoral immunity* involves the production of antibodies by specialized cells
   b. *Cell-mediated immunity* involves highly specialized cells that carry out defensive activities directly
2. Two important aspects of the body's specific defense mechanisms are the ability to recognize antigens and memorize them
   a. Lymphocytes have a particular type of receptor (which differs from lymphocyte to lymphocyte) protruding from their cell membrane, enabling them to recognize antigens
      (1) This receptor is specific to a certain antigen

(2) When an antigen passes through a lymph node (which contains large numbers of lymphocytes), it binds to the lymphocyte (either a T cell or B cell or both) that has the appropriate receptor

(3) Primed by this interaction with the antigen, the selected lymphocytes begin to grow and divide, producing effector cells (T cells, plasma cells, and memory cells) that are specialized for defending against the invading antigen

b. Memory cells store (memorize) information about the antigen; they can survive for decades, whereas the T cells and plasma cells produced during this interchange may last only a few days

3. The body's response to antigen exposure can be one of two types —a primary response or a secondary response —depending on whether memory cells are present for a particular antigen

a. A *primary response* takes place when no memory cells are available

(1) This response occurs when the body is exposed to an antigen for the first time

(2) The primary response is slow; it may take several days before the B cells can produce plasma cells, which in turn produce antibodies specific to the antigen

b. A *secondary response* occurs when memory cells are already present

(1) This response results from previous exposure to a particular antigen

(2) The secondary response is swift because memory cells can proliferate rapidly to produce plasma cells and effector cells specific to the antigen

4. Vaccinations can help the body become immune to specific antigens

a. A vaccination presents the body with an inactive or weakened form of the pathogen, which cannot cause the disease itself

b. The body launches a primary immune response against the antigen, resulting in the production of memory cells

c. If the body is exposed to the same pathogen introduced by the vaccination, a swift secondary response is initiated

5. The type of immunity achieved by antigen exposure can be either active or passive

a. *Active immunity* (achieved through a primary or secondary response) occurs when antigens enter the body naturally

b. *Passive immunity* results when the body receives antibodies produced by another body (for example, when antibodies cross the placenta from mother to fetus)

6. Five classes of antibodies, or ***immunoglobulins*** *(Ig),* have been identified based on their structure and function

a. *IgM,* the largest of the immunoglobulins, is produced early in the body's immune response; it causes agglutination (clumping) of antigens and helps to activate complement

b. *IgG,* the most abundant of the circulating antibodies, can cross the placenta; usually produced after an infection is well established, it can activate complement

c. *IgA* is found in body secretions, such as sweat, saliva, and tears; the primary antibody of colostrum (the initial secretion from a mother's breast after birth), IgA can bind with the invading substance and carry it out of the body through various secretions

d. *IgD* binds to the plasma membrane of B cells; it probably acts as the antigen receptor for B cells

e. *IgE* is associated with allergic responses

7. Within each immunoglobulin class is an infinite variety of antibodies, each specific to a particular antigen
   a. Antibodies alone usually do not directly destroy the invading antigen but instead trigger various *effector mechanisms*
   b. Each effector mechanism is activated by the selective binding of the antigen to the antibody, a process that results in an *antigen-antibody complex*
   c. The antigen-antibody complex causes neutralization, agglutination, precipitation, or activation of the complement system
      (1) *Neutralization* of the antigen is the simplest effector mechanism; the antibody binds to the antigen and neutralizes its harmful effects, allowing phagocytic cells to dispose of it
      (2) In *agglutination,* a single antibody binds to more than one invading cell, causing them to clump together; the clumped cells are then easily destroyed by phagocytes
      (3) *Precipitation* is similar to agglutination; the antibody binds to antigen molecules (rather than cells) and forms an immobile precipitate destroyed by phagocytes
      (4) *Activation* of the complement system excites the inflammatory response and leads to cell lysis
8. In contrast to B cells, T cells (the main agents of cellular immunity) cannot be activated by free antigens present in body fluids; T-cell activation begins when an antigen-processing cell, such as a macrophage, ingests a foreign object
   a. This antigen-processing cell then attaches part of the ingested object to its cell membrane (this is called the *antigenic determinant)*
   b. In combination with the major histocompatibility complex (MHC) protein already present on the surface of the cell, a receptor is created that can bind a circulating T cell
   c. When the T cell binds to the receptor, a response is initiated
   d. The response involves the release of cytokines (***interleukin I*** and interleukin II) and the activation of other T cells
      (1) Interleukin I, released by the macrophage, stimulates the growth and cell division of the T cell
      (2) Interleukin II, released by the activated T cell, stimulates helper T cells (which stimulate B cells to grow) and increases the production of cytotoxic T cells
      (3) At the end of the response, suppressor T cells release a cytokine that inhibits the activity of B cells and other T cells

Study Activities

1. Outline the functions of the major body systems in vertebrates.
2. Make a representative drawing of the human heart, including all chambers and valves, and indicate the direction of blood movement.

3. Describe the regulation of respiration in mammals.
4. Describe the mechanical and chemical breakdown of food as it passes through the alimentary canal.
5. Compare the effects of the sympathetic and parasympathetic nervous systems.
6. Trace the movement of sodium and potassium ions across the nerve cell membrane during impulse transmission.
7. Draw and label the structures in the outer, middle, and inner ear.
8. Define the sliding filament model of muscle contraction.
9. Differentiate between humoral and cell-mediated immunity.
10. Describe how vaccination provides immunity against specific antigens.
11. Identify the structure and function of the five classes of immunoglobulins.
12. List the possible effects of the antigen-antibody complex on the complement system.

Appendices

A: Geologic Timetable

B: Taxonomic Classification of Living Organisms

C: Glossary

Selected References

Index

Appendix A: Geologic Timetable

ERA	PERIOD	EPOCH	YEARS AGO*	SIGNIFICANT EVENT
Precambrian			4,600	Origin of earth
			3,500	Prokaryotes appear
			2,500	Photosynthesis occurs
			1,500	Eukaryotes appear
			700	Animals appear
Paleozoic	Cambrian		600	Invertebrates appear; algae diversify
	Ordovician		500	First vertebrates (jawless fishes) appear
	Silurian		450	Vascular plants and arthropods colonize land
	Devonian		400	Amphibians and insects appear
	Carboniferous		360	First seed plants and reptiles appear
	Permian		280	Most modern insects appear
Mesozoic	Triassic		250	Dinosaurs, mammals, and birds appear
	Jurassic		215	Gymnosperms and dinosaurs dominate land
	Cretaceous		144	Angiosperms appear; dinosaurs become extinct
Cenozoic	Paleogene	Paleocene	65	Major radiation of mammals, birds, and pollinating insects occurs
		Eocene	54	Angiosperms dominate land
		Oligocene	38	Most modern mammals (including apes) appear
	Neogene	Miocene	24	Radiation of mammals and angiosperms continues
		Pliocene	5	Apelike human ancestors appear
		Pleistocene	1.8	Ice ages occur; humans appear
		Recent	0.01	Modern civilizations arise

* In millions of years; times are approximate.

Appendix B: Taxonomic Classification of Living Organisms

MONERA (KINGDOM)

- Archaebacteria **(Major Group)**
- Eubacteria **(Major Group)**

Subgroup
- Cyanobacteria
- Phototrophic bacteria
- Pseudomonads
- Spirochetes
- Endospore-forming bacteria
- Enteric bacteria
- Rickettsiae & Chlamydiae
- Mycoplasmas
- Actinomycetes
- Myxobacteria

PROTISTA (KINGDOM)

Phylum*

- Protozoa **(Major Group)**
  - Rhizopoda
  - Actinopoda
  - Foraminifera
  - Apicomplexa
  - Zoomastigina
  - Ciliophora
- Algal Protists **(Major Group)**
  - Dinoflagellata
  - Chrysophyta
  - Bacillariophyta
  - Euglenophyta
  - Chlorophyta
  - Phaeophyta
  - Rhodophyta
- Fungus-like Protists **(Major Group)**
  - Myxomycota
  - Acrasiomycota
  - Oomycota

FUNGI (KINGDOM)

- Zygomycota
- Ascomycota
- Basidiomycota
- Deuteromycota

PLANTAE (KINGDOM)

- Nonvascular **(Major Group)** — Bryophyta

*In the Fungi and Plantae Kingdoms, the term *Division* is often substituted for the term *Phylum*.

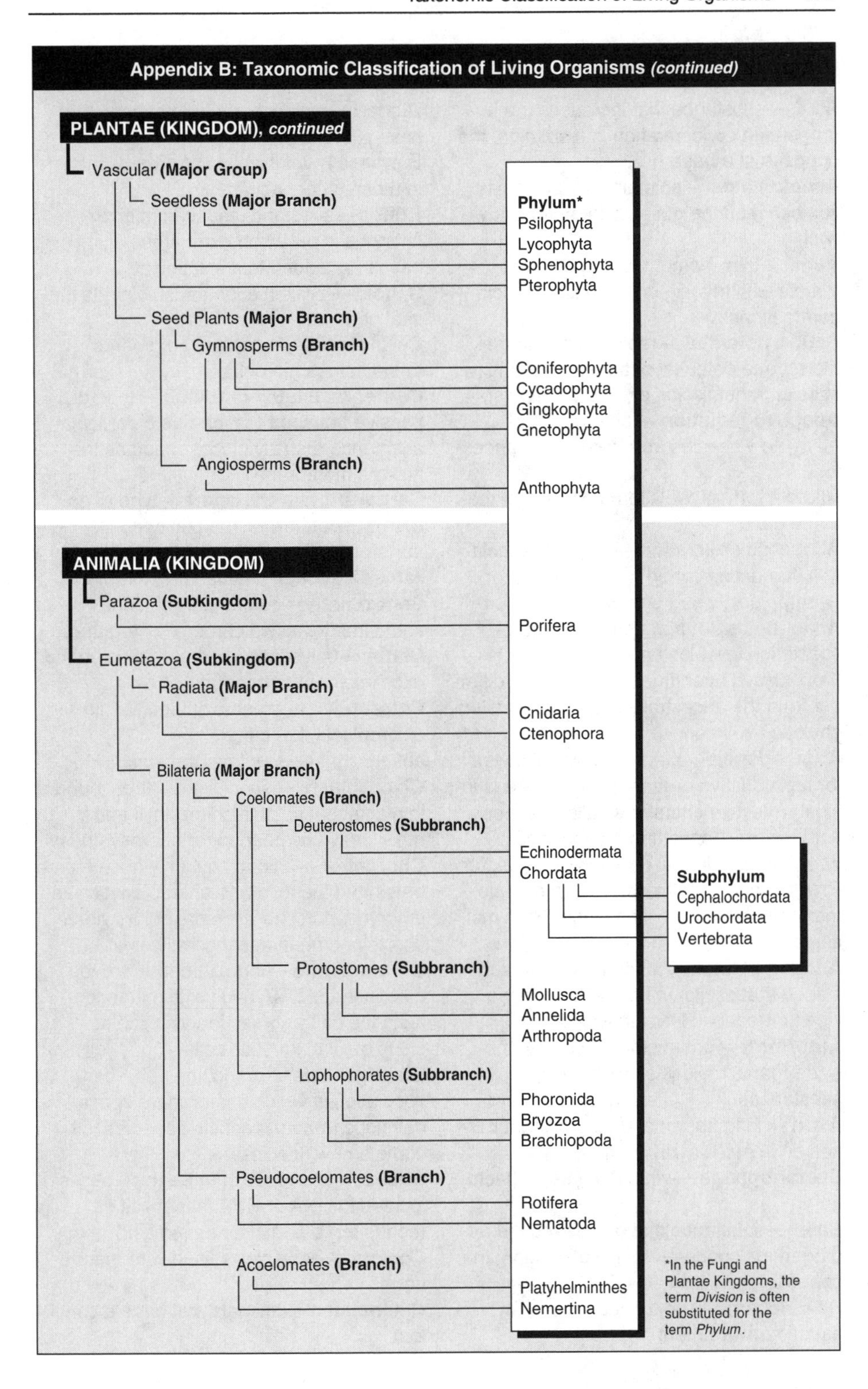
Appendix B: Taxonomic Classification of Living Organisms (continued)
PLANTAE (KINGDOM), continued
Vascular (Major Group)
Seedless (Major Branch)
Phylum*
Psilophyta
Lycophyta
Sphenophyta
Pterophyta
Seed Plants (Major Branch)
Gymnosperms (Branch)
Coniferophyta
Cycadophyta
Gingkophyta
Gnetophyta
Angiosperms (Branch)
Anthophyta
ANIMALIA (KINGDOM)
Parazoa (Subkingdom)
Porifera
Eumetazoa (Subkingdom)
Radiata (Major Branch)
Cnidaria
Ctenophora
Bilateria (Major Branch)
Coelomates (Branch)
Deuterostomes (Subbranch)
Echinodermata
Chordata
Subphylum
Cephalochordata
Urochordata
Vertebrata
Protostomes (Subbranch)
Mollusca
Annelida
Arthropoda
Lophophorates (Subbranch)
Phoronida
Bryozoa
Brachiopoda
Pseudocoelomates (Branch)
Rotifera
Nematoda
Acoelomates (Branch)
Platyhelminthes
Nemertina
*In the Fungi and Plantae Kingdoms, the term Division is often substituted for the term Phylum.

Appendix C: Glossary

Acid —substance that increases the hydrogen ion concentration in a solution; the opposite of a base
Acoelomates —animals that lack cavities between the gut and the outer body wall
Actin —thin, filamentous protein that interacts with myosin in producing muscle contractions
Action potential —rapid change in the membrane potential of a neuron that results in transmission of a nerve impulse
Adaptive radiation —emergence of many new species from a common ancestor
Allele —any of the alternate forms of the same gene
Allopatric speciation —mode of speciation that occurs when a physical barrier geographically segregates a population
Anagenesis —unbranched pattern of speciation in which an entire species is transformed over time into a species different from the ancestral species; also called phyletic evolution
Angiosperms —plants that have flowers for reproductive organs and form seeds inside protective chambers called ovaries
Anticodon —specialized sequence of three nucleotides on one end of a transfer RNA molecule that recognizes a complementary codon on a messenger RNA molecule
Antigen —foreign, macromolecular substance that elicits an immune response when introduced into a host's body
Autotroph —organism that can synthesize organic molecules from inorganic substances
Axon —long neuronal projection that carries an impulse away from the cell
Bacteriophage —virus that infects bacteria
Base —substance that decreases the hydrogen ion concentration in a solution; the opposite of an acid
Benthic zone —bottom surface of an aquatic environment
Bilateria —branch of eumetazoans that have bilateral symmetry
Biomass —total weight of organic matter in a particular habitat
Buffer —substance that can minimize changes in pH when extraneous acids or bases are added to a solution
Capsid —protein coat that surrounds the viral genome
Carotenoids —accessory pigments found in the chloroplast
Carrier-facilitated diffusion —type of passive transport that carries a particular substance (usually glucose) across the plasma membrane
Carrier-protein transport —type of active transport in which an enzyme-like carrier protein binds to a molecule and carries it through the plasma membrane
Centromere —central part of a chromosome that joins two chromatids together
Chitin —polysacchride that makes up the exoskeleton of arthropods
Chlorophyll —green pigments within chloroplasts that can absorb light energy from the sun
Chloroplasts —special structures found in plants that contain chlorophyll and other enzymes needed for photosynthesis
Chordates —animal phylum whose members have the following characteristics as embryos: notochord, dorsal hollow nerve chord, and pharyngeal gill slits
Chromosome —long, thread-like body composed of DNA and protein that contains the cell's genes; found in the nucleus of all eukaryotic cells
Cladogenesis —branching pattern of speciation in which one or more species develops from an ancestral species to become a new species
Codon —sequence of three nucleotides located in DNA or RNA that specifies (codes for) a particular amino acid
Coelom —body cavity lined with mesoderm
Coelomates —animals that have a coelom

Companion cell —type of plant cell whose nucleus and ribosomes serve a sieve tube cell
Cuticle —waxy covering on the surface of terrestrial plant stems and leaves that prevents dessication
Cytoplasm —living matter found between the cell membrane and the nucleus
Cytoplasmic streaming —phenomenon in plant cells in which the entire cytoplasm flows around the cell in the space between the vacuole and the plasma membrane to speed up the distribution of materials within the cell
Dendrite —one of numerous, highly branched neuronal projections that carry impulses toward a nerve cell body
Deuterostomes —organisms characterized by radial and indeterminate cleavage and a mouth that does not arise from the blastopore
Dicot —subdivision of flowering plants whose members have two embryonic seed leaves (cotyledons)
Diffusion —molecular movement of particles from a more concentrated area to a less concentrated area
Ecology —study of the interactions of organisms with each other and with their environment
Endoskeleton —hard, rigid structures within the soft tissues of an animal
Epiphyte —plant that grows on the surface of another plant for support
Eumetazoa —subkingdom that includes all animals except sponges
Evolution —all changes that have transformed life on earth
Exoskeleton —hard substance that surrounds an animal; provides protection and a place for the attachment of muscles
Fertilization —fusion of two haploid gametes to produce a diploid zygote
Fibrin —activated form of the blood-clotting protein fibrinogen
Food chain —transfer of food from one trophic level to the next
Food web —interconnected feeding relationships within an ecosystem
Fossil —preserved remnant or impression of a living organism from a past gene pool
Gametes —haploid egg or sperm cells that fuse during sexual reproduction to produce a diploid zygote
Gametophyte —multicellular haploid form occurring in organisms that undergo alternation of generations; it mitotically produces haploid gametes that fuse and grow into the sporophyte generation
Genome —genetic material of an organism
Genotype —genetic composition of an organism
Gymnosperms —plants that do not have enclosed chambers for seeds to develop; reproductive structures are cones
Hemoglobin —iron-containing protein found in red blood cells that binds oxygen
Heterotroph —organism that cannot manufacture organic compounds and must feed on other organisms to obtain energy
Hierarchy —arrangement into a graded or ranked series
Hominid —bipedal primate mammals comprising modern humans and their recent ancestors
Homologous chromosomes —chromosome pairs that possess genes for the same trait at similar loci
Hormone —circulating chemical signal in multicellular organisms that interacts with target cells to produce a specific effect
Hyphae —filaments that make up the body of a fungus
Immunoglobulins —class of proteins that make up antibodies
Interferon —chemical substance produced by the immune system that helps cells resist viruses
Interleukin —cytokine chemical regulator released by the immune system that causes immune system cells to grow and reproduce
Invertebrates —animals without backbones

Lignin —primary noncellulose component of plant tissue that provides structural support
Lysogenic cycle —method of viral reproduction in which the virus's genome is incorporated into the host cell's DNA
Lytic cycle —method of viral reproduction that kills the host cell
Macroevolution —evolutionary change involving many populations over a long period that results in new designs, evolutionary trends, adaptive radiation, or mass extinctions
Microevolution —change in a population's gene pool over a period of time
Mitochondria —special structures found in most cells that contain the enzymes needed for aerobic respiration
Molting —periodic shedding of the exoskeleton by arthropods to permit growth
Monocot —subdivision of flowering plants whose members have one embryonic seed leaf (cotyledon)
Mutation —change in the DNA of the genes
Mycelium —branched network of hyphae found in fungi
Myosin —thick, filamentous protein that interacts with actin in muscle contractions
Natural selection —nonrandom reproduction of different phenotypes that results from the interaction of organisms with their environment
Neurotransmitter —chemical substance that transmits nerve impulses across a synapse
Nitrogen fixation —assimilation of atmospheric nitrogen by certain organisms into nitrogenous compounds that can be used by plants
Notochord —flexible, longitudinal rod formed from dorsal mesoderm; found in all chordate embryos, it is located between the gut and the nerve chord
Omnivores —animals that eat both plants and animals
Ontogeny —embyronic development of an organism
Osmosis —movement of water across a selectively permeable membrane
Oviparous —method of reproduction in which offspring hatch from eggs laid outside the mother's body
Ovoviviparous —method of reproduction in which offspring hatch from eggs retained within the mother's uterus
Parasites —organisms that obtain their nutrients from the body fluids of living hosts
Parazoa —subkingdom of animals consisting of the sponges
Phenotype —expressed traits of an organism
Pheromones —volatile chemicals that serve as communication signals between animals
Photosynthesis —process in which energy, light, and chlorophyll interact to manufacture glucose from carbon dioxide and water
pH scale —measure of hydrogen ion concentration ranging from 0 to 14; a pH level less than 7 is acidic, one greater than 7 is basic (or alkaline)
Phylogeny —evolutionary history of a species or group of related species
Phytoplankton —passively floating plant life in a body of water
Plasmodesmata —open channels in the cell walls of plants through which cytoplasm connects adjacent cells
Polar transport —rapid, unidirectional transport of substances from one plant cell to the next
Pollination —process by which pollen is placed on the stigma of a plant; it is a prerequisite to fertilization
Population —group of individuals of one species that live in a particular geographic area
Primary consumers —organisms that eat plants or algae (herbivores)
Productive cycle —method of viral replication in which the virus's genome is incorporated into a host cell's DNA; the mature virus exits the host cell by budding
Prophage —genome of a bacteriophage inserted into a host cell's DNA

Protostomes —organisms characterized by spiral, determinate cleavage and development of the blastopore into the mouth
Provirus —viral genome after insertion into a host cell's DNA
Pseudocoelomates —animals whose body cavities are not completely lined by mesoderm
Pseudopodia —cellular extensions of amoeboid cells used in feeding and movement
Radiata —animals with radial symmetry
Respiration —anaerobic or aerobic process by which energy is liberated from glucose
Retrovirus —RNA virus that reproduces by transcribing its RNA into DNA and inserting this DNA into a host cell's DNA
Reverse transcriptase —enzyme found in some RNA viruses that uses RNA as a template for DNA synthesis
Sarcomere —basic, repeating unit of striated muscle
Sarcoplasmic reticulum —modified form of endoplasmic reticulum in striated muscle that stores calcium
Secondary consumers —carnivores that eat herbivores
Seed —plant embryo and a store of food packaged within a resistant coat
Sodium-potassium pump —protein in the plasma membrane of animal cells that actively transports sodium out of and potassium into the cell
Sorus —cluster of sporangia on a fern leaf
Species —population or group of populations whose members possess similar anatomic characteristics and can interbreed with each other
Spore —meiotically produced haploid cell that divides mitotically to produce a multicellular individual, the gametophyte
Sporophyte —multicellular diploid form occurring in organisms that undergo alternation of generations; it meiotically produces haploid spores that grow into the gametophyte generation
Stomata —pores surrounded by guard cells that permit the exchange of gases between a plant and its external environment
Succession —transition in the composition of species within a biologic community
Symbiosis —dependent relationship between organisms of two different species
Sympatric speciation —mode of speciation that occurs when a subpopulation is reproductively isolated from its parent population
Synapse —space between two adjacent neurons
Taxonomy —branch of science charged with naming and classifying living things
Trophic level —feeding level of an ecosystem that determines the route of energy flow and the cycling of chemicals
Vertebrates —animals with backbones, including mammals, birds, reptiles, amphibians, and most fishes
Villi —fingerlike projections of the small intestine that increase its surface area
Virion —complete virus particle
Virus —noncellular obligatory parasite consisting of a nucleic acid surrounded by a protein coat
Viviparous —method of reproduction in which offspring are born alive after developing within the mother

Selected References

Camp, P.S., and Arms, K. *Exploring Biology,* 3rd ed. Philadelphia: W.B. Saunders Co., 1987.

Campbell, N. *Biology,* 2nd ed. Redwood City, Calif.: Benjamin-Cummings Publishing Co., 1990.

Curtis, H., and Barnes, N.S. *Biology,* 5th ed. New York: Worth Publishers, 1989.

Hopson, J.L., and Wessells, N.K. *Essentials of Biology.* New York: McGraw-Hill Publishing Co., 1990.

Keeton, W.T., and Gould, J.L. *Biological Science,* 4th ed. New York: W.W. Norton & Co., 1987.

Mader, S. *Human Biology,* 3rd ed. Dubuque, Iowa: W.C. Brown & Co., 1990.

Mason, W., and Marshall, N.L. *Perspectives in Biology,* 2nd ed. Dubuque, Iowa: Kendall-Hunt Publishing Co., 1990.

Rabe, F.W. *Introduction to Biology,* 2nd ed. Dubuque, Iowa: Kendall-Hunt Publishing Co., 1991.

Raven, and Johnson. *Understanding Biology,* 2nd ed. St. Louis: C.V. Mosby Co., 1989.

Wessells, N., and Hopson, J. *Biology.* New York: Random House, 1988.

Index

i refers to an illustration; t, to a table

D

E

i refers to an illustration; t, to a table

i refers to an illustration; t, to a table

Q,R

S

i refers to an illustration; t, to a table

T

U

V

W

X,Y,Z

i refers to an illustration; t, to a table